SECOND EDITION

A PHOTOGRAPHIC
ATLAS
OF THE HUMAN BODY

With Selected Cat, Sheep, and Cow Dissections

GERARD J. TORTORA

Bergen Community College

WILEY

JOHN WILEY & SONS, INC.

Executive Editor	Bonnie Roesch
Marketing Manager	Clay Stone
Project Editor	Mary O'Sullivan
Associate Production Manager	Kelly Tavares
Senior Designer	Karin Gerdes Kincheloe
Art Coordinator	Claudia Durrell
Photo Manager	Hilary Newman

This book was typeset by Progressive Information Technologies. This book was printed and bound by Von Hoffmann Press, Inc. The cover was printed by Von Hoffmann Press, Inc.

The paper in this book was manufactured by a mill whose forest management programs include sustained yield harvesting of its timberlands. Sustained yield harvesting principles ensure that the number of trees cut each year does not exceed the amount of new growth.

This book is printed on acid-free paper ♾

To order books or for customer service please, call 1(800)-CALL-WILEY (225-5945).

Library of Congress Cataloging in Publication Data:
Tortora, Gerard J.
   A photographic atlas of the human body : with selected cat, sheep, and cow dissections /
Gerard J. Tortora.
     p. cm.
  ISBN 0-471-42064-6
   1. Human anatomy Atlases. 2. Cats—Dissection. 3. Sheep—Dissection. 4. Cows—Dissection. I. Title.
  [DNLM: 1. Anatomy Atlases. 2. Dissection Atlases. 3. Physiology Atlases. / QS 17 T712p 2000]
  QM25.T67   2000
  611'.022'2—dc21
  DNLM/DLC
  for Library of Congress
                                      99-32135
                                          CIP

Printed in the United States of America.

10 9 8 7 6 5

Table of Contents

Preface

A Photographic Atlas of the Human Body with Selected Cat, Sheep, and Cow Dissections, 2e, is designed to accompany any textbook of anatomy or anatomy and physiology and may be used in conjunction with or in lieu of a laboratory manual.

The study of the gross anatomical features of the human body is enhanced by the use of a photographic atlas to supplement your experience in the dissecting room or in those courses that do not include actual dissection. The clearly labeled cadaver photographs in this atlas were provided mainly by Mark Nielsen, of the University of Utah. They are organized by body system, and provide you with a stunning, visual reference to gross anatomy. The cadaver photos are supplemented by histological aspects of various organ and surface anatomy photos. You will benefit further from the helpful orientation diagrams, which accompany many of the photographs.

As you will see from the table of contents, this atlas covers all of the topics discussed in a typical anatomy or a combined anatomy and physiology course. The atlas begins with anatomical orientation and tissues, then progresses through each organ system and ends with developmental biology. Where comparisons are helpful sheep and cow dissection photos appear with their human counterpart. A complete set of cat dissection photographs appears in an appendix.

Each photomicrograph in this atlas is accompanied by a line diagram and an inset that indicates a primary location of the tissue in the body. An icon with each photograph indicates whether it is a light or an electron micrograph and gives its magnification.

A c k n o w l e d g m e n t s

There are many people whom I wish to thank for their efforts in helping me prepare this atlas. Bonnie Roesch, Executive Editor, my editor for so many of my publications, continues to provide the guidance, creativity, and professionalism which have impacted so profoundly on all of my books. There is no way that I can adequately express my gratitude to Bonnie. Clay Stone, Marketing Manager, has been a vital component of all of my publications with Wiley. Clay coordinates a network of individuals who present the salient features of my books to professors. The feedback that Clay provides me from professors and students is invaluable. Kelly Tavares, Associate Production Manager, has once again distinguished herself as a "super" coordinator for all aspects of the production process. Her commitment to and passion for her job is so obvious and appreciated. Karin Gerdes Kincheloe, Senior Designer, is responsible for the cover and text design. Throughout, the pages are visually attractive, pedagogically effective, and student-friendly. For so many years, Claudia Durrell has been my art coordinator. Her organizational skills and artistic vision continue to amaze me. She always knows what to do and she does it so well. Hilary Newman, the Photo Manager, has pursued my photo requests with tenacity and has provided me with many options from which to choose. I also want to acknowledge Mary O'Sullivan, Project Editor, for her outstanding efforts in coordinating the development of supplements for my text, such as this atlas.

In preparing this photographic atlas, I have strived to keep uppermost in my mind your needs, as a student. As always, I could benefit from your comments and suggestions, which I hope you will send to me at the address below.

Gerard J. Tortora
Science and Technology, S229
Bergen Community College
400 Paramus Road
Paramus, NJ 07652

TABLE 1.1 | *Directional Terms*

Directional Term	Definition	Example of use
Superior (soo'-PEER-ē-or) (**cephalic** or **cranial**)	Toward the head or the upper part of a structure.	The heart is superior to the liver.
Inferior (in'-FEER-ē-or) (**caudal**)	Away from the head or the lower part of a structure.	The stomach is inferior to the lungs.
Anterior (an-TEER-ē-or) (**ventral**)[*]	Nearer to or at the front of the body.	The sternum (breastbone) is anterior to the heart.
Posterior (pos-TEER-ē-or) (**dorsal**)[*]	Nearer to or at the back of the body.	The esophagus is posterior to the trachea (windpipe).
Medial (MĒ-dē-al)	Nearer to the midline[†].	The ulna is medial to the radius.
Lateral (LAT-er-al)	Farther from the midline.	The lungs are lateral to the heart.
Intermediate (in'-ter-MĒ-dē-at)	Between two structures.	The transverse colon is intermediate between the ascending and descending colons.
Ipsilateral (ip-si-LAT-er-al)	On the same side of the midline.	The gallbladder and ascending colon are ipsilateral.
Contralateral (CON-tra-lat-er-al)	On the opposite side of the midline.	The ascending and descending colons are contralateral.
Proximal (PROK-si-mal)	Nearer to the attachment of a limb to the trunk; nearer to the origination of a structure.	The humerus is proximal to the radius.
Distal (DIS-tal)	Farther from the attachment of a limb to the trunk; farther from the origination of a structure.	The phalanges are distal to the carpals.
Superficial (soo'-per-FISH-al)	Toward or on the surface of the body.	The ribs are superficial to the lungs.
Deep (DĒP)	Away from the surface of the body.	The ribs are deep to the skin of the chest and back.

[*]In four-legged animals anterior = cephalic (toward the head), ventral = inferior, posterior = caudal (toward the tail), and dorsal = superior.

[†]The midline is an imaginary vertical line that divides the body into equal right and left sides.

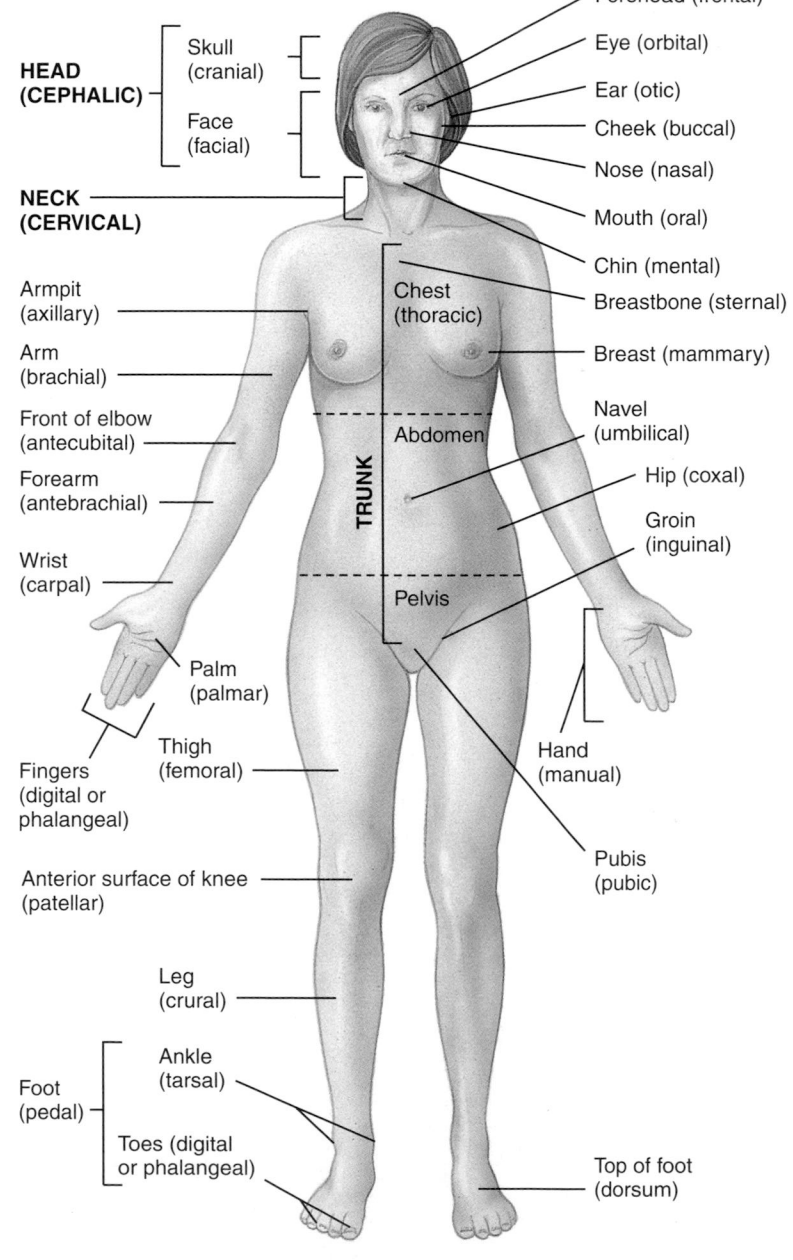

HEAD (CEPHALIC)
- Skull (cranial)
- Face (facial)

NECK (CERVICAL)

- Forehead (frontal)
- Eye (orbital)
- Ear (otic)
- Cheek (buccal)
- Nose (nasal)
- Mouth (oral)
- Chin (mental)
- Breastbone (sternal)
- Breast (mammary)
- Navel (umbilical)
- Hip (coxal)
- Groin (inguinal)

Armpit (axillary)

Arm (brachial)

Front of elbow (antecubital)

Forearm (antebrachial)

Wrist (carpal)

TRUNK

Chest (thoracic)

Abdomen

Pelvis

Palm (palmar)

Thigh (femoral)

Fingers (digital or phalangeal)

Anterior surface of knee (patellar)

Leg (crural)

Foot (pedal)

Ankle (tarsal)

Toes (digital or phalangeal)

Hand (manual)

Pubis (pubic)

Top of foot (dorsum)

Anterior view

FIGURE 1.1

The anatomical position and regional names. In the anatomical position, the subject stands erect facing the observer with the head level and eyes facing forward. The feet are flat on the floor and directed forward, and the arms are at the sides with the palms turned forward. The common names and anatomical terms, in parentheses, are indicated for many regions of the body.

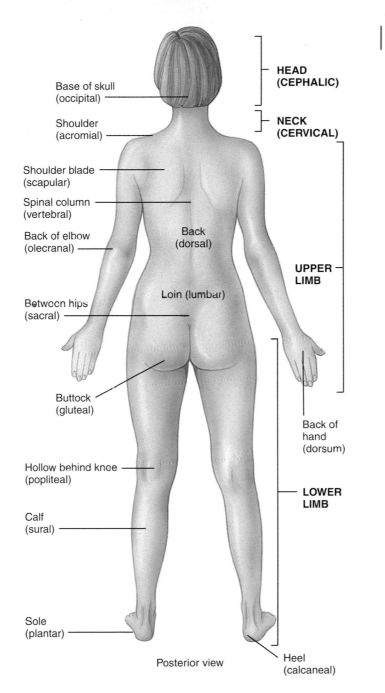

FIGURE 1.2
The anatomical position and regional names

Base of skull
(occipital)

HEAD
(CEPHALIC)

Shoulder
(acromial)

NECK
(CERVICAL)

Shoulder blade
(scapular)

Spinal column
(vertebral)

Back of elbow
(olecranal)

Back
(dorsal)

UPPER
LIMB

Between hips
(sacral)

Loin (lumbar)

Buttock
(gluteal)

Back of
hand
(dorsum)

Hollow behind knee
(popliteal)

LOWER
LIMB

Calf
(sural)

Sole
(plantar)

Posterior view

Heel
(calcaneal)

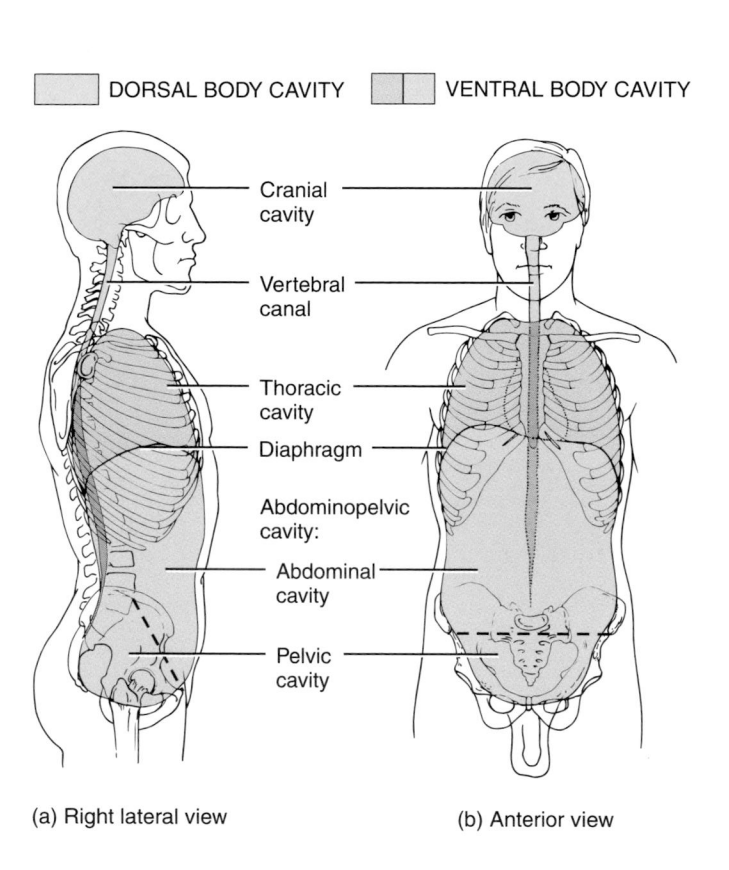

DORSAL BODY CAVITY   VENTRAL BODY CAVITY

Cranial cavity

Vertebral canal

Thoracic cavity

Diaphragm

Abdominopelvic cavity:

Abdominal cavity

Pelvic cavity

(a) Right lateral view   (b) Anterior view

CAVITY	COMMENTS
DORSAL	
Cranial	Formed by cranial bones and contains brain and its coverings.
Vertebral	Formed by vertebral column and contains spinal cord and the beginnings of spinal nerves.
VENTRAL	
Thoracic	Chest cavity; separated from abdominal cavity by diaphragm.
Pleural (right and left)	Contain lungs.
Pericardial	Contains heart.
Mediastinum	Region between the lungs from the breastbone to backbone that contains heart, thymus, esophagus, trachea, bronchi, and several large blood and lymphatic vessels.
Abdominopelvic	Subdivided into abdominal and pelvic cavities.
Abdominal	Contains stomach, spleen, liver, gallbladder, small intestine, and most of large intestine.
Pelvic	Contains urinary bladder, portions of the large intestine, and internal reproductive organs.

FIGURE 1.3
Principal subdivisions of the dorsal and ventral body cavities

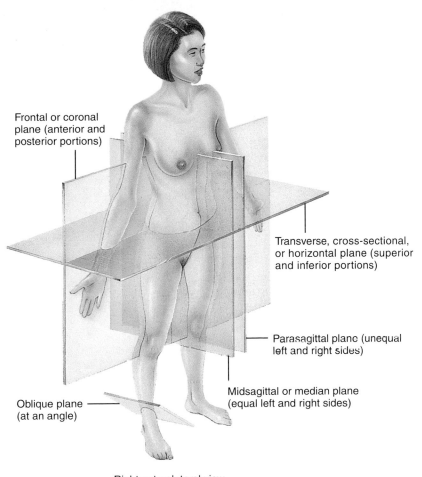

Frontal or coronal plane (anterior and posterior portions)

Transverse, cross-sectional, or horizontal plane (superior and inferior portions)

Parasagittal plane (unequal left and right sides)

Midsagittal or median plane (equal left and right sides)

Oblique plane (at an angle)

Right anterolateral view

FIGURE 1.4

Planes are imaginary flat surfaces that divide the entire body or individual organs into various portions. The descriptions in parentheses indicate how each plane divides the body.

(a) Transverse plane

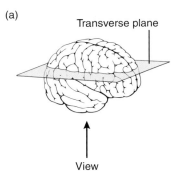 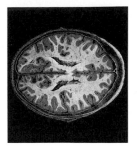

View

Transverse (cross) section

(b) Frontal plane

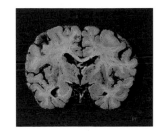

View

Frontal section

(c) Midsagittal plane

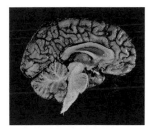

View

Midsagittal section

FIGURE 1.5

Planes and sections. The planes are shown in the diagrams on the left and the sections that result are shown in the photographs of the brain on the right.

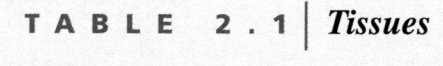

TABLE 2.1 | *Tissues*

Tissue	Comment	Tissue	Comment
Epithelial Tissue		**Connective Tissue, continued**	
I. Covering and lining	Forms outer covering of body and some viscera; lines body cavities, some viscera, blood vessels, and ducts; makes up parts of sense organs.	B. Dense	
		Dense regular	Forms tendons and ligaments.
		Dense irregular	Found in dermis of skin, fasciae, and membranes around various structures.
A. Simple	Single layer of cells.	Elastic	Provides stretch and strength.
Squamous	Flat, scalelike cells.	C. Cartilage	Has no blood or nerve supply.
Cuboidal	Cube-shaped cells.	Hyaline	Found at ends of long bones and ribs.
Columnar	Rectangular-shaped cells.	Fibrocartilage	Found in pubic symphysis and intervertebral discs.
Pseudostratified	Single layer of cells that appears to be stratified.	Elastic	Found in larynx and external ear.
B. Stratified	Two or more layers of cells.	D. Bone (osseous)	Contains very rigid intercelluar substance and is classified as compact or spongy.
Squamous	Flat, scalelike cells in apical (superficial) layer.		
Cuboidal	Cube-shaped cells in apical layer.	E. Blood (vascular)	Liquid connective tissue consisting of blood plasma and formed elements (red blood cells, white blood cells, and platelets).
Columnar	Rectangular-shaped cells in apical layer.		
Transitional	Cells variable in shape.		
II. Glandular	Forms secretory portions of glands.		
A. Exocrine	Secrete products into ducts.	**Muscular Tissue**	Highly specialized for contraction.
B. Endocrine	Secrete hormones into interstitial fluid and then into the blood.	I. Skeletal	Usually attached to bones, striated, voluntary.
		II. Cardiac	Found in the heart, striated, involuntary.
Connective Tissue		III. Smooth (visceral)	Found in viscera and blood vessels, nonstriated, involuntary.
I. Embryonic	Present primarily in embryo and fetus.		
A. Mesenchyme	Embryonic tissue from which all other connective tissues develop.	**Nervous Tissue**	
B. Mucous	Fetal tissue found in umbilical cord.	I. Neurons	Specialized for detecting stimuli, converting them into action potentials, and conducting action potentials.
III Mature	Found in newborn.		
A. Loose		II. Neuroglia	Protect and support neurons.
Areolar	One of the most abundant connective tissues.		
Adipose	Specialized for fat storage.		
Reticular	Forms stroma (framework) of certain organs.		

FIGURE 2.1
Histology of epithelial tissue

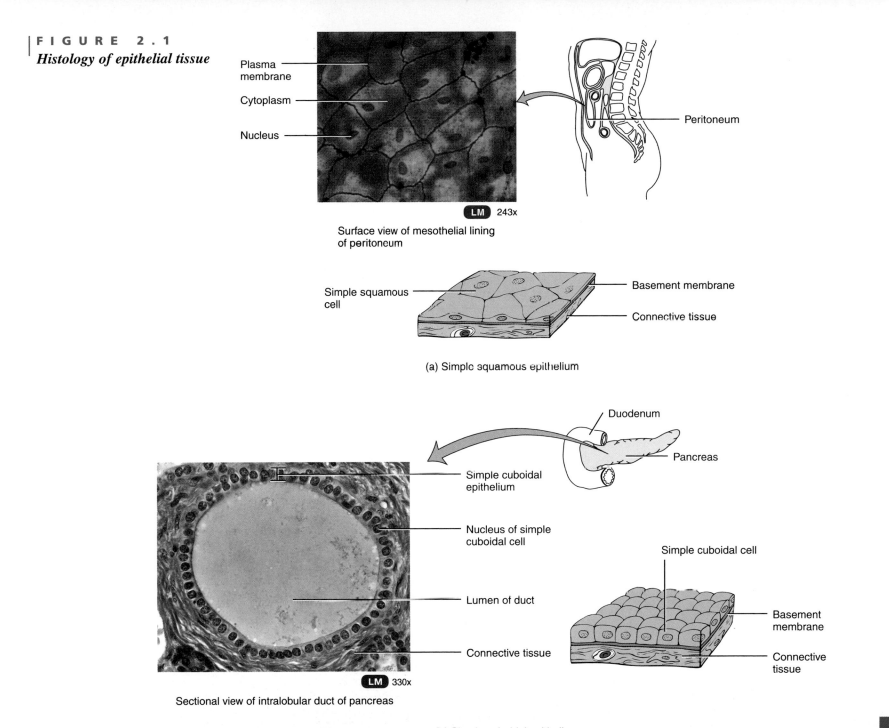

Plasma membrane

Cytoplasm

Nucleus

Peritoneum

LM 243x

Surface view of mesothelial lining of peritoneum

Simple squamous cell

Basement membrane

Connective tissue

(a) Simple squamous epithelium

Duodenum

Pancreas

Simple cuboidal epithelium

Nucleus of simple cuboidal cell

Simple cuboidal cell

Lumen of duct

Basement membrane

Connective tissue

Connective tissue

LM 330x

Sectional view of intralobular duct of pancreas

(b) Simple cuboidal epithelium

FIGURE 2.1
Histology of epithelial tissue, continued

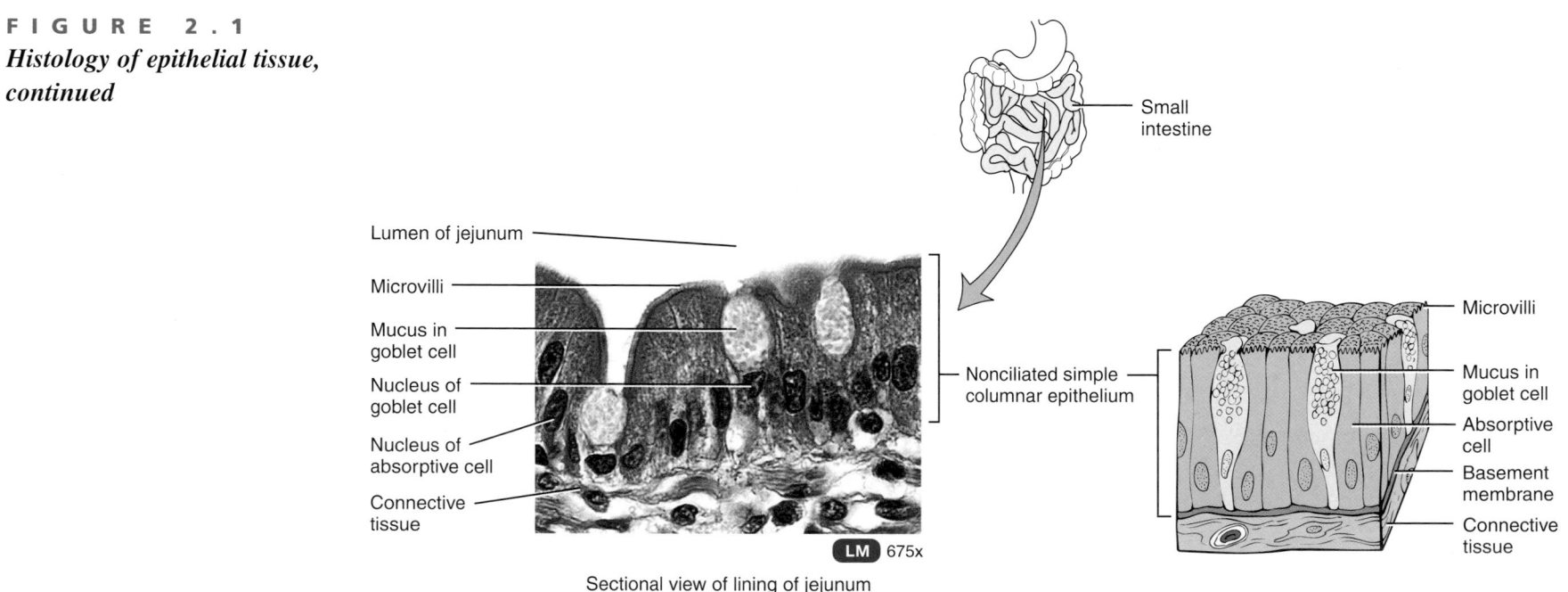

Small intestine

Lumen of jejunum

Microvilli

Mucus in goblet cell

Nucleus of goblet cell

Nucleus of absorptive cell

Connective tissue

Nonciliated simple columnar epithelium

LM 675x

Sectional view of lining of jejunum of small intestine

Microvilli

Mucus in goblet cell

Absorptive cell

Basement membrane

Connective tissue

(c) Nonciliated simple columnar epithelium

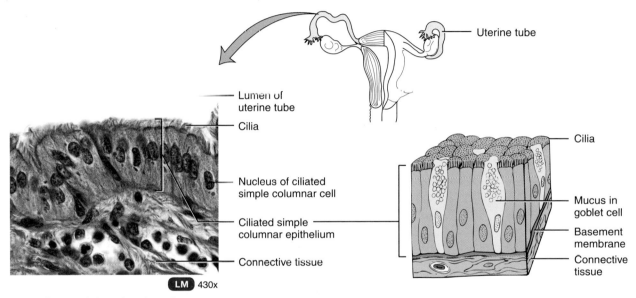

Lumen of
uterine tube

Cilia

Nucleus of ciliated
simple columnar cell

Ciliated simple
columnar epithelium

Connective tissue

Uterine tube

Cilia

Mucus in
goblet cell

Basement
membrane

Connective
tissue

LM 430x

Sectional view of uterine tube

(d) Ciliated simple columnar epithelium

FIGURE 2.1
Histology of epithelial tissue, continued

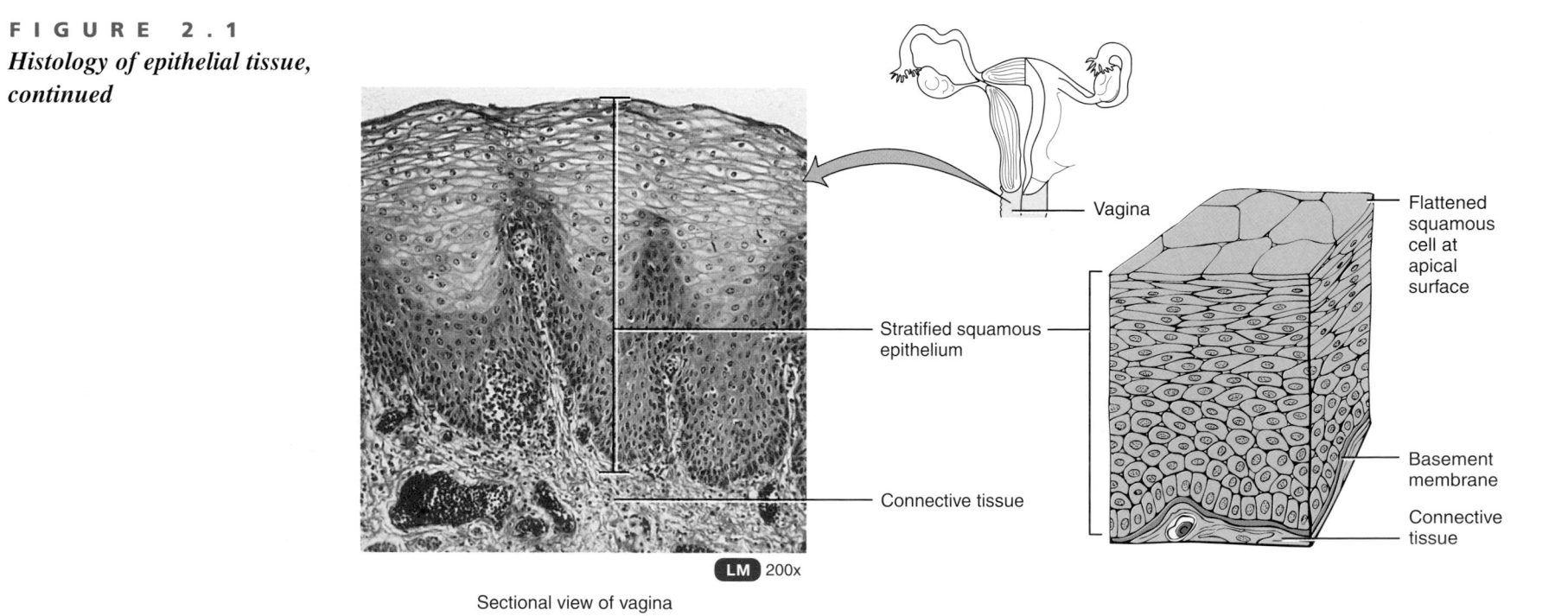

Vagina

Stratified squamous epithelium

Connective tissue

LM 200x

Sectional view of vagina

Flattened squamous cell at apical surface

Basement membrane

Connective tissue

(e) Stratified squamous epithelium

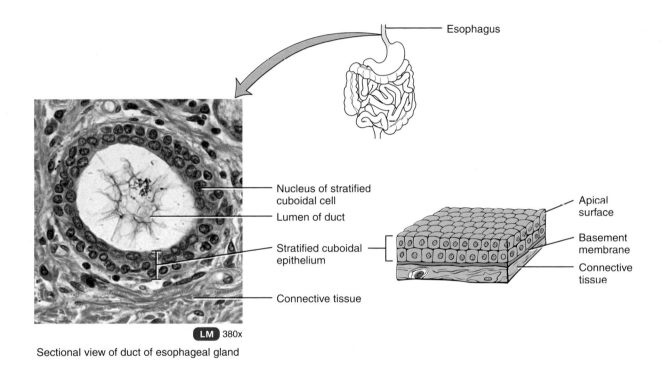

Esophagus

Nucleus of stratified cuboidal cell

Lumen of duct

Stratified cuboidal epithelium

Connective tissue

Apical surface

Basement membrane

Connective tissue

LM 380x

Sectional view of duct of esophageal gland

(f) Stratified cuboidal epithelium

FIGURE 2.1
Histology of epithelial tissue, continued

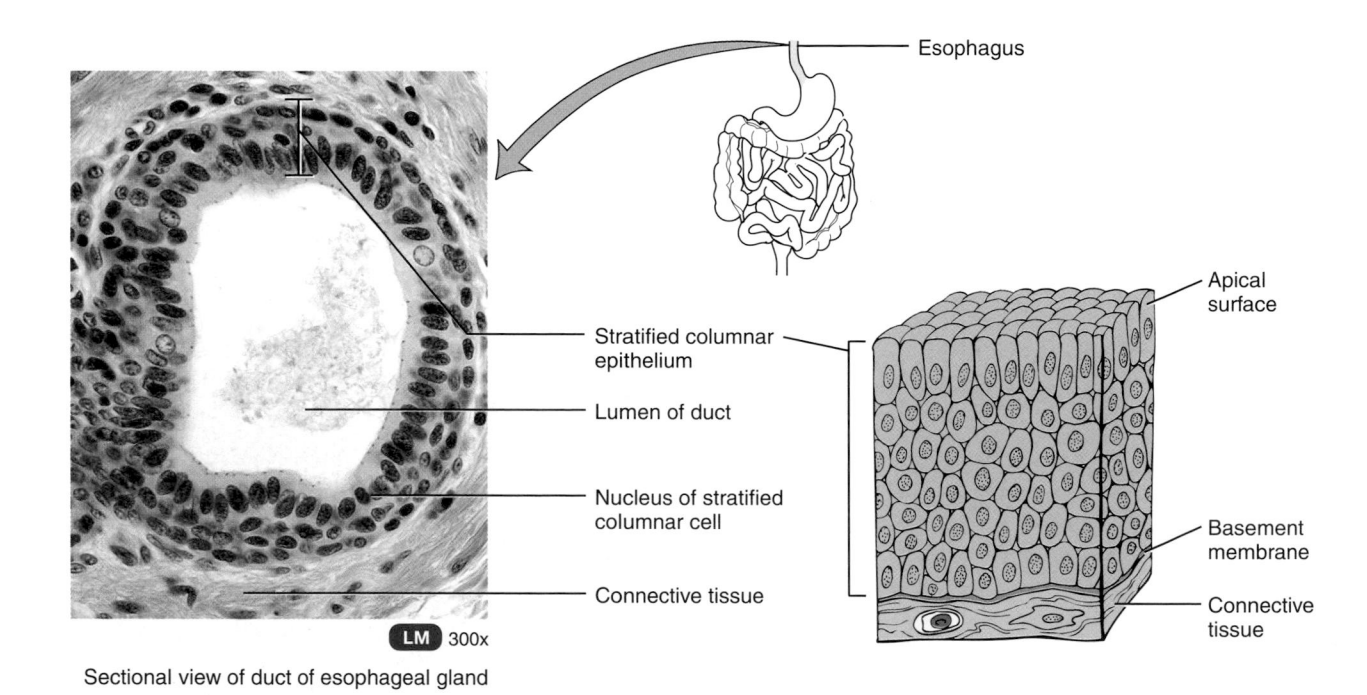

Esophagus

Stratified columnar epithelium

Lumen of duct

Nucleus of stratified columnar cell

Connective tissue

LM 300x

Sectional view of duct of esophageal gland

Apical surface

Basement membrane

Connective tissue

(g) Stratified columnar epithelium

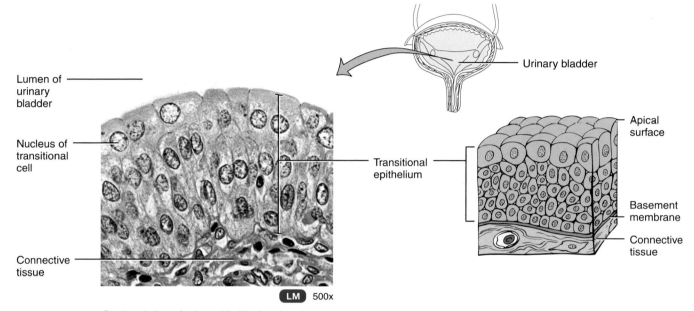

Lumen of urinary bladder

Nucleus of transitional cell

Connective tissue

LM 500x

Sectional view of urinary bladder in relaxed state

Urinary bladder

Transitional epithelium

Apical surface

Basement membrane

Connective tissue

(h) Relaxed transitional epithelium

FIGURE 2.1
Histology of epithelial tissue, continued

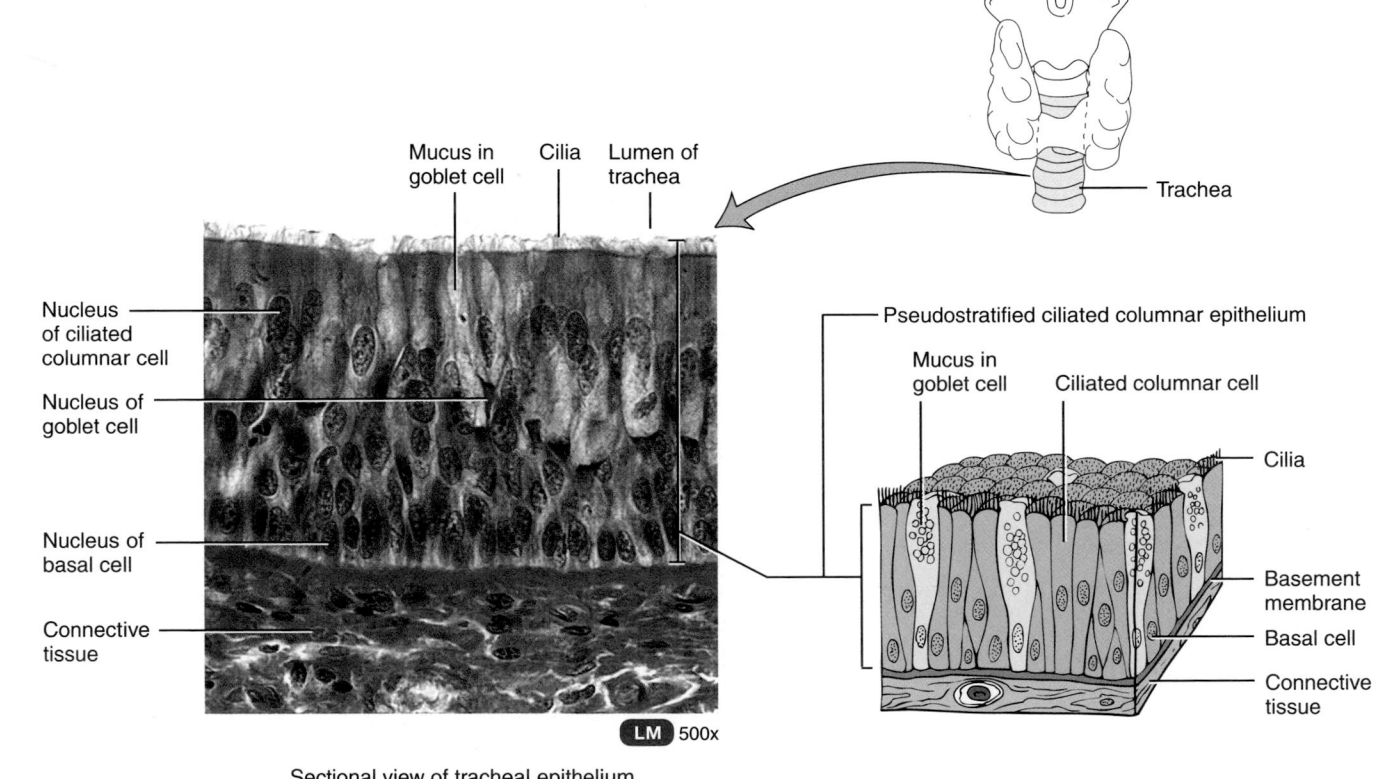

Mucus in goblet cell

Cilia

Lumen of trachea

Trachea

Nucleus of ciliated columnar cell

Nucleus of goblet cell

Nucleus of basal cell

Connective tissue

Pseudostratified ciliated columnar epithelium

Mucus in goblet cell

Ciliated columnar cell

Cilia

Basement membrane

Basal cell

Connective tissue

LM 500x

Sectional view of tracheal epithelium

(i) Pseudostratified ciliated columnar epithelium

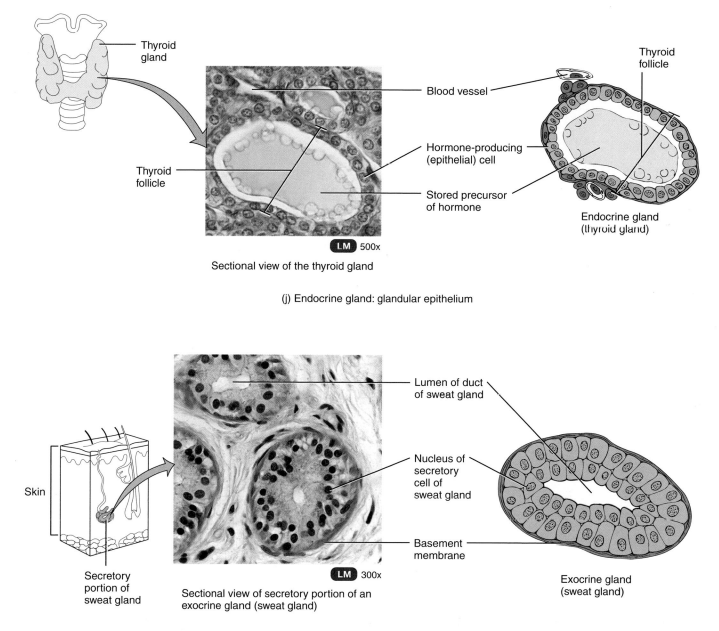

Thyroid gland

Blood vessel

Thyroid follicle

Thyroid follicle

Hormone-producing (epithelial) cell

Stored precursor of hormone

Endocrine gland (thyroid gland)

LM 500x

Sectional view of the thyroid gland

(j) Endocrine gland: glandular epithelium

Skin

Secretory portion of sweat gland

Lumen of duct of sweat gland

Nucleus of secretory cell of sweat gland

Basement membrane

Exocrine gland (sweat gland)

LM 300x

Sectional view of secretory portion of an exocrine gland (sweat gland)

(k) Exocrine gland: glandular epithelium

FIGURE 2.2
Histology of connective tissue

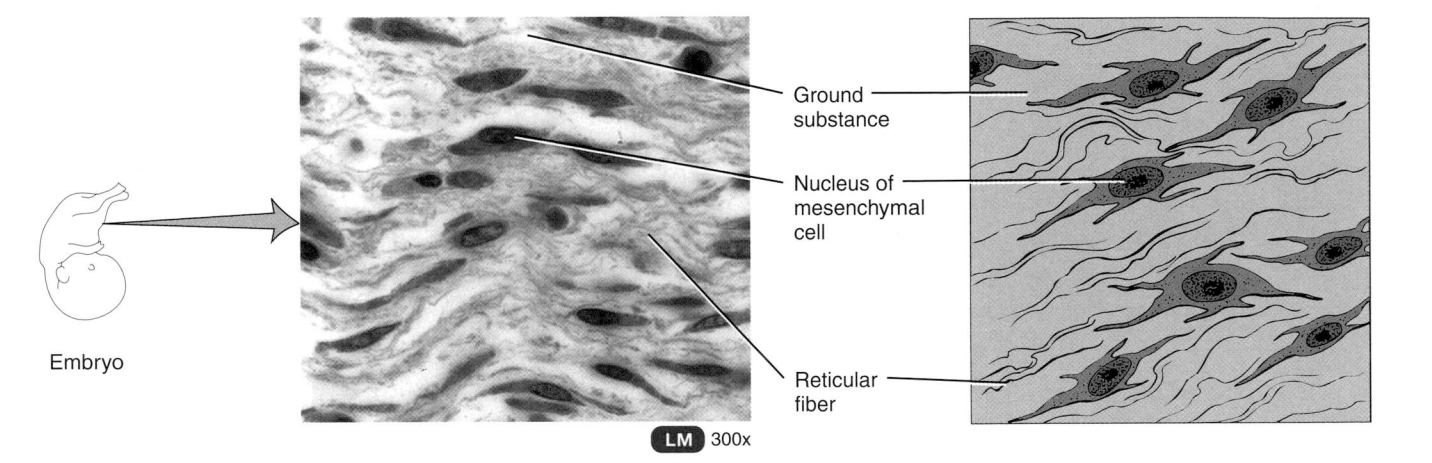

Embryo

Ground
substance

Nucleus of
mesenchymal
cell

Reticular
fiber

LM 300x

Sectional view of a developing embryo

(a) Mesenchyme

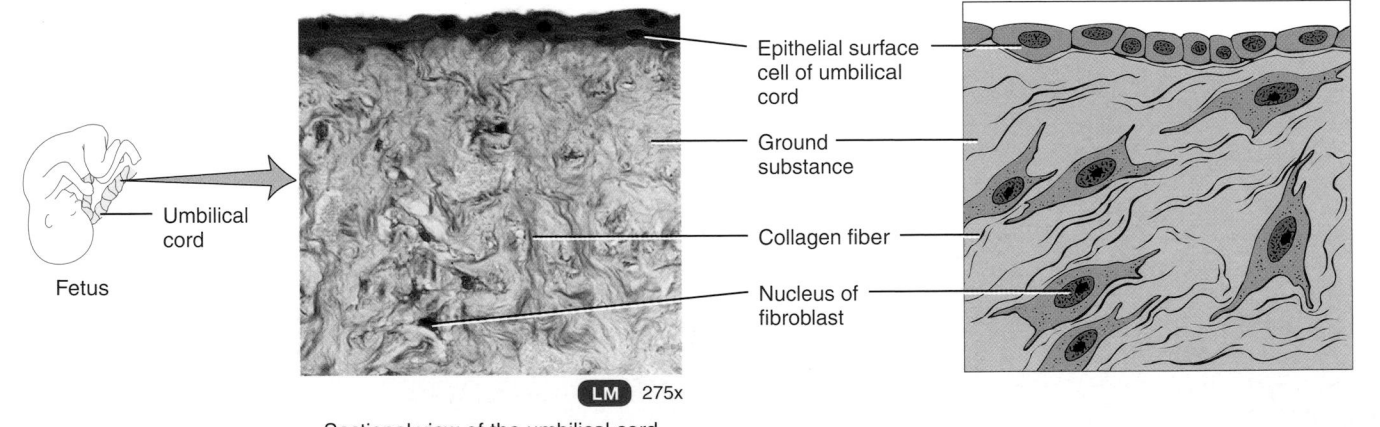

Fetus

Umbilical
cord

Epithelial surface
cell of umbilical
cord

Ground
substance

Collagen fiber

Nucleus of
fibroblast

LM 275x

Sectional view of the umbilical cord

(b) Mucous connective tissue

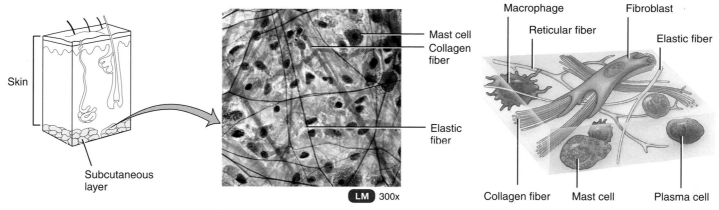

Mast cell
Collagen fiber
Elastic fiber

Macrophage
Reticular fiber
Fibroblast
Elastic fiber

LM 300x

Sectional view of subcutaneous areolar connective tissue

Skin

Subcutaneous layer

Collagen fiber
Mast cell
Plasma cell

(c) Areolar connective tissue

Heart

Fat

Nucleus of adipocyte
Cytoplasm
Fat-storage area of adipocyte
Blood vessel
Plasma membrane

LM 300x

Sectional view of adipose tissue showing adipocytes of white fat

(d) Adipose tissue

FIGURE 2.2

Histology of connective tissue, continued

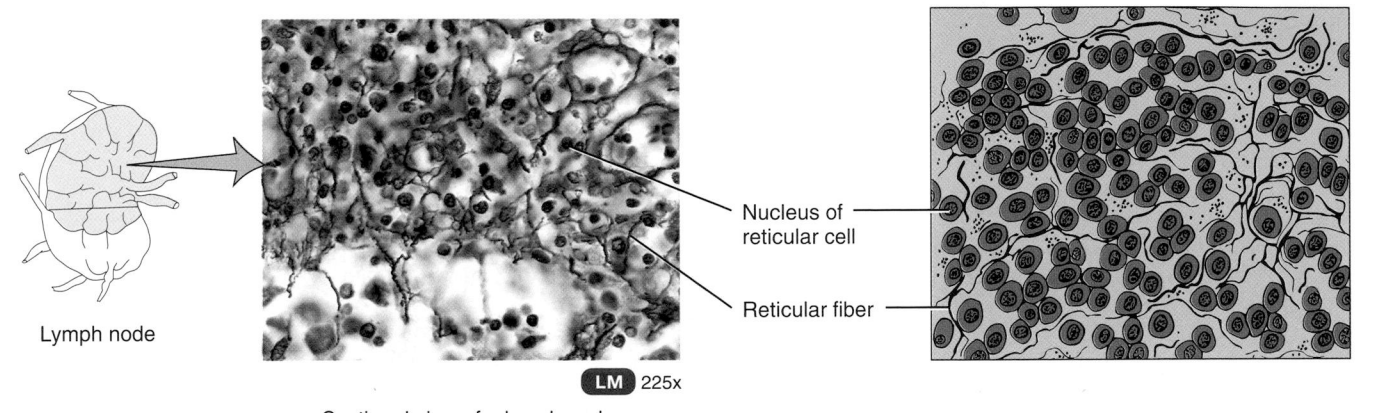

Lymph node

Nucleus of
reticular cell

Reticular fiber

LM 225x

Sectional view of a lymph node

(e) Reticular connective tissue

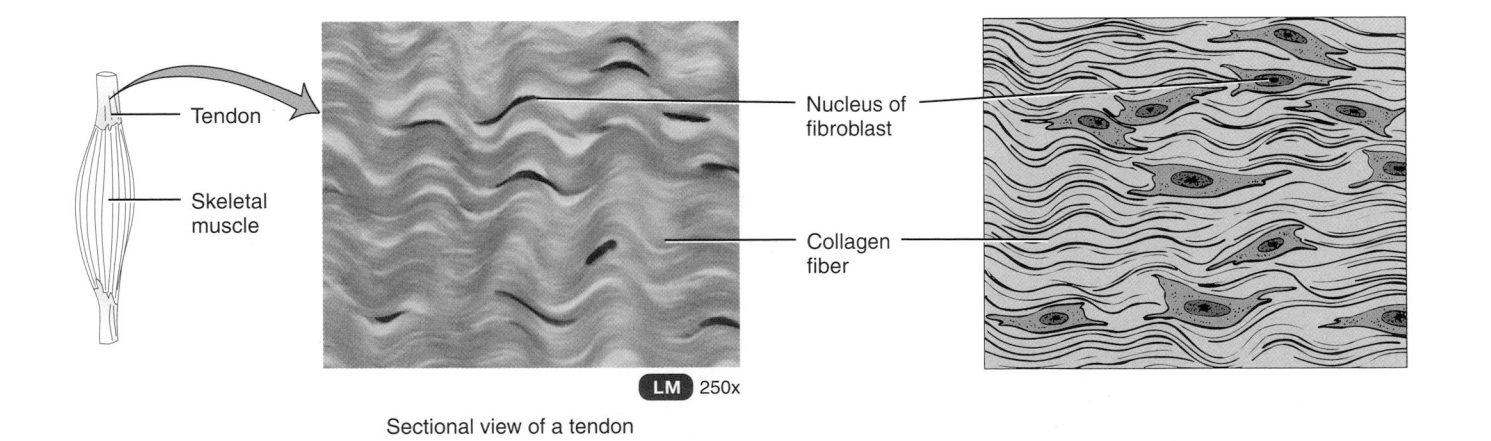

Tendon

Skeletal
muscle

Nucleus of
fibroblast

Collagen
fiber

LM 250x

Sectional view of a tendon

(f) Dense regular connective tissue

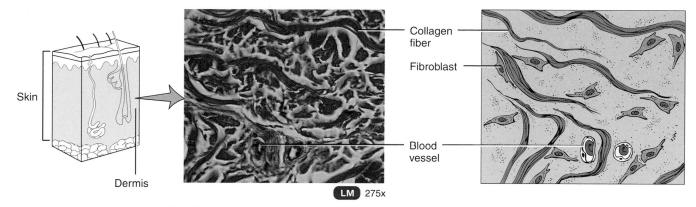

Skin

Dermis

Collagen
fiber

Fibroblast

Blood
vessel

LM 275x

Sectional view of reticular region of dermis of skin

(g) Dense irregular connective tissue

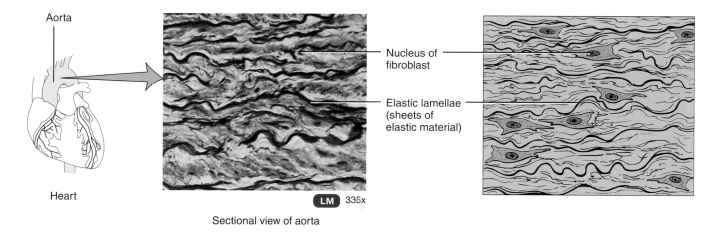

Aorta

Heart

Nucleus of
fibroblast

Elastic lamellae
(sheets of
elastic material)

LM 335x

Sectional view of aorta

(h) Elastic connective tissue

FIGURE 2.2
Histology of connective tissue, continued

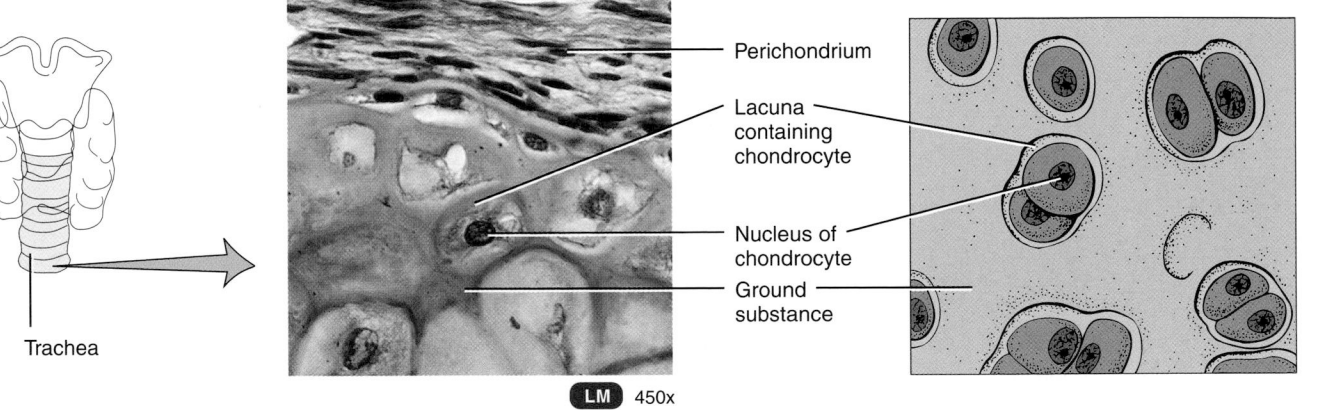

Trachea

Perichondrium

Lacuna containing chondrocyte

Nucleus of chondrocyte

Ground substance

LM 450x

Sectional view from trachea

(i) Hyaline cartilage

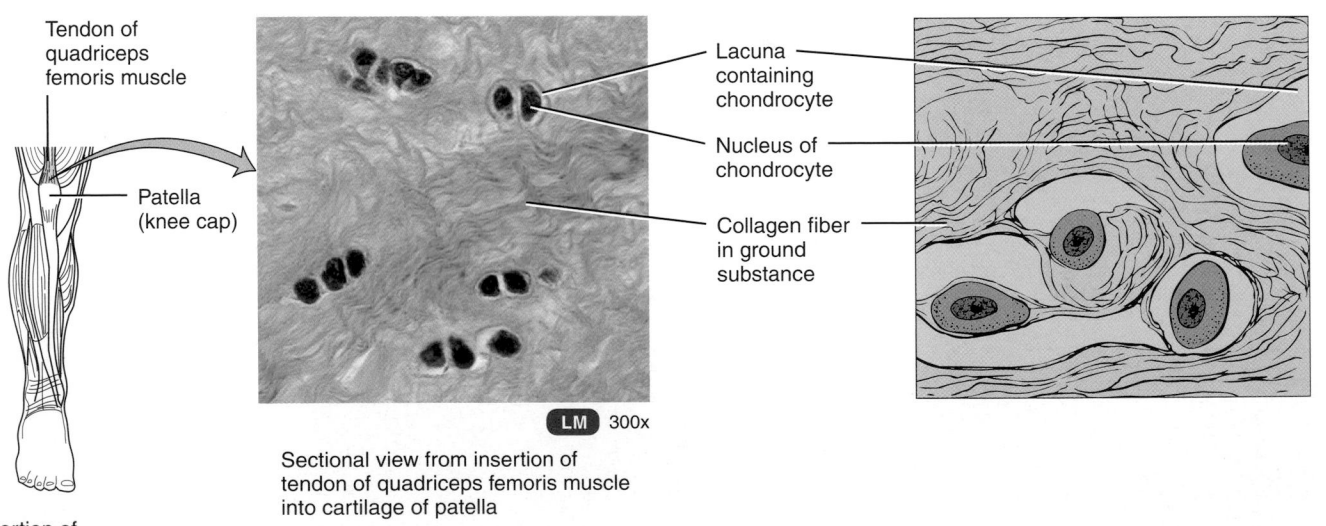

Tendon of quadriceps femoris muscle

Patella (knee cap)

Portion of right lower limb

Lacuna containing chondrocyte

Nucleus of chondrocyte

Collagen fiber in ground substance

LM 300x

Sectional view from insertion of tendon of quadriceps femoris muscle into cartilage of patella

(j) Fibrocartilage

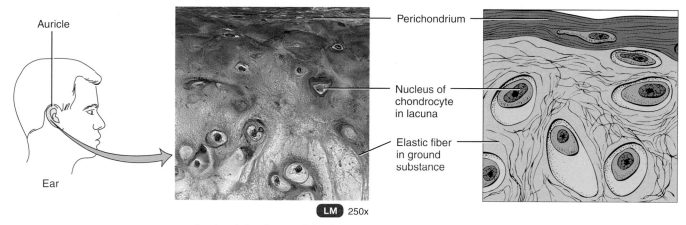

Auricle

Ear

Perichondrium

Nucleus of chondrocyte in lacuna

Elastic fiber in ground substance

LM 250x

Sectional view from auricle of ear

(k) Elastic cartilage

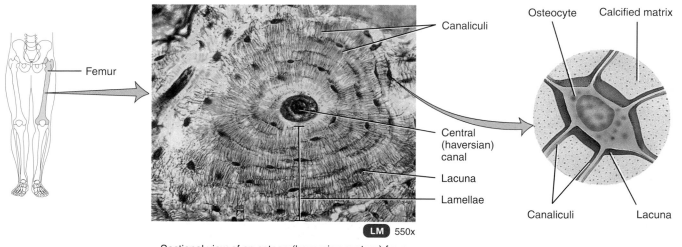

Femur

Canaliculi

Central (haversian) canal

Lacuna

Lamellae

LM 550x

Sectional view of an osteon (haversian system) from femur (thigh bone)

Osteocyte   Calcified matrix

Canaliculi   Lacuna

(l) Bone tissue

F I G U R E   2 . 2

Histology of connective tissue, continued

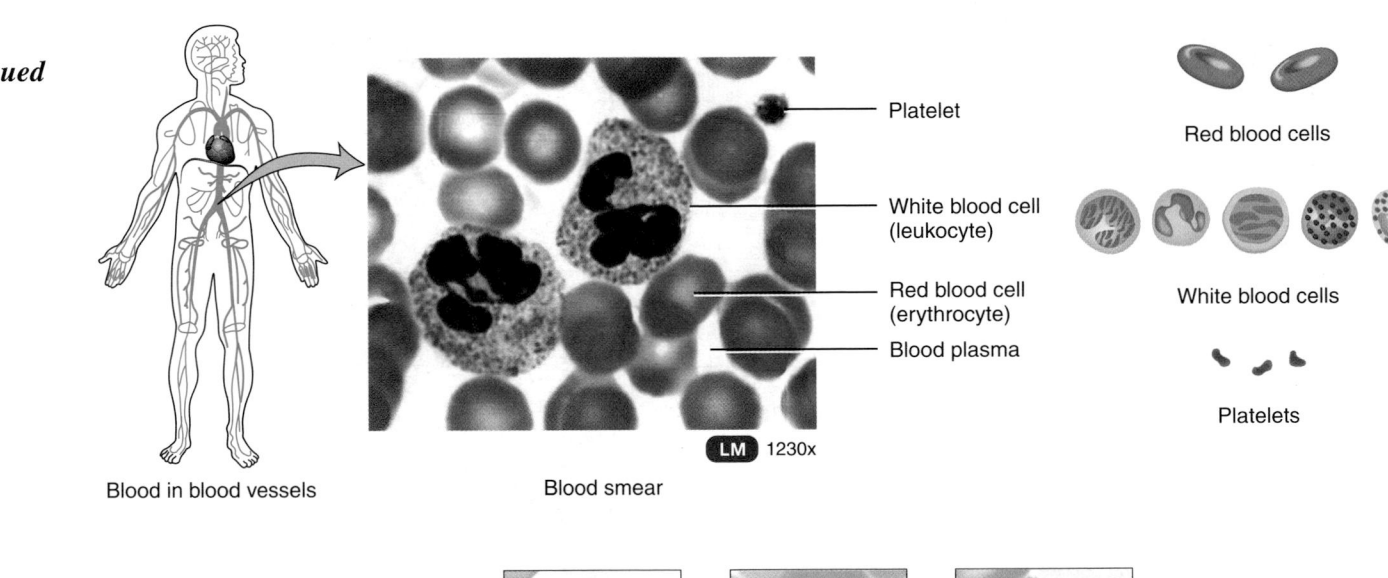

Platelet

White blood cell
(leukocyte)

Red blood cell
(erythrocyte)

Blood plasma

LM 1230x

Blood in blood vessels

Blood smear

Red blood cells

White blood cells

Platelets

Eosinophil

Basophil

Neutrophil

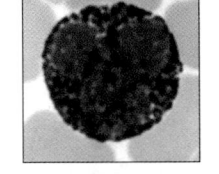

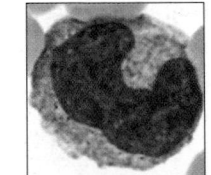

All LM 1420x

Small lymphocyte

Monocyte

Details of a blood smear

(m) Blood tissue

FIGURE 2.3
Histology of muscular tissue

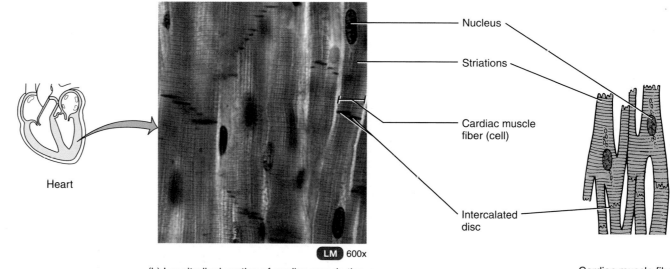

Skeletal muscle

Skeletal muscle fiber (cell)

Nucleus

Striations

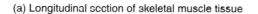
LM 400x

Skeletal muscle fiber

(a) Longitudinal section of skeletal muscle tissue

Heart

Nucleus

Striations

Cardiac muscle fiber (cell)

Intercalated disc

LM 600x

(b) Longitudinal section of cardiac muscle tissue

Cardiac muscle fibers

FIGURE 2.3
Histology of muscular tissue, continued

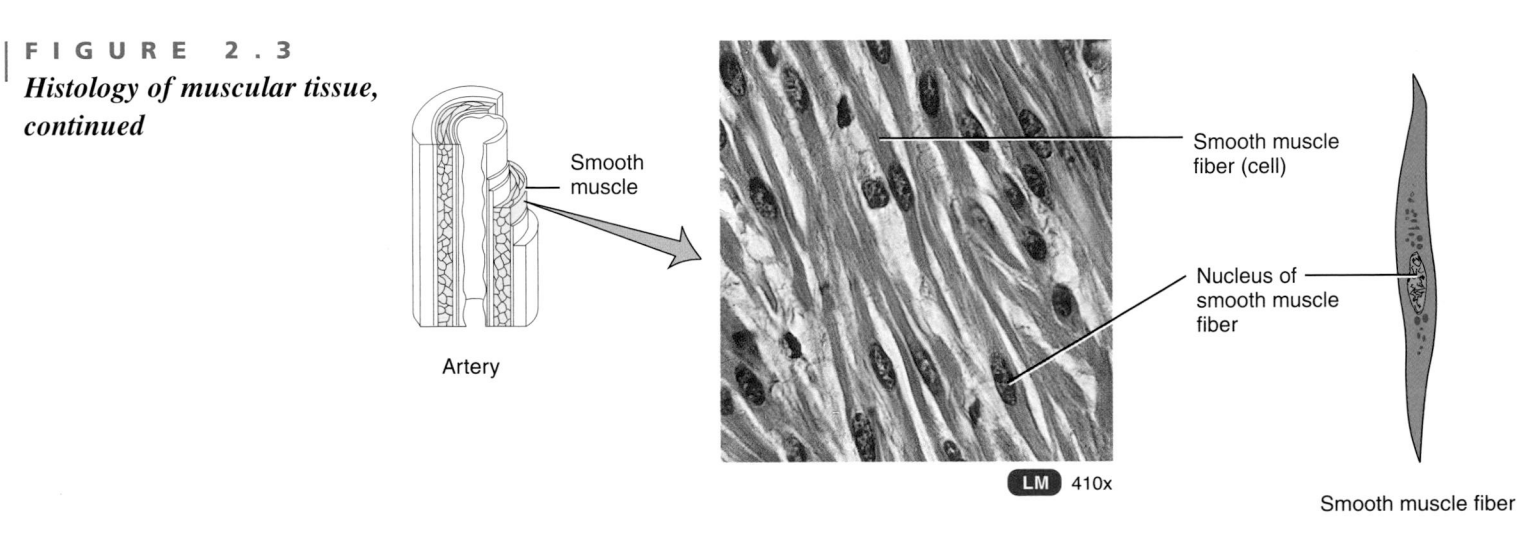

Smooth muscle

Artery

Smooth muscle fiber (cell)

Nucleus of smooth muscle fiber

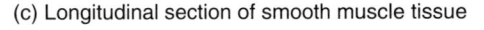 LM 410x

Smooth muscle fiber

(c) Longitudinal section of smooth muscle tissue

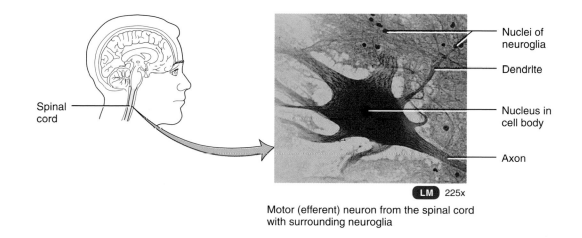

Spinal cord

Nuclei of neuroglia

Dendrlte

Nucleus in cell body

Axon

LM 225x

Motor (efferent) neuron from the spinal cord with surrounding neuroglia

F I G U R E   2 . 4
Histology of nervous tissue and neuroglia

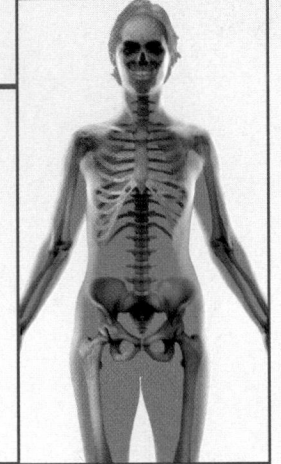

TABLE 3.1 | *Divisions of the Adult Skeletal System*

Regions of the Skeleton	Number of Bones	Regions of the Skeleton	Number of Bones
Axial Skeleton		**Appendicular Skeleton**	
Skull		Pectoral (shoulder) girdles	
Cranium	8	*Clavicle*	2
Face	14	*Scapula*	2
Hyoid	1	Upper limbs (extremities)	
Auditory ossicles	6	*Humerus*	2
Vertebral column	26	*Ulna*	2
Thorax		*Radius*	2
Sternum	1	*Carpals*	16
Ribs	24	*Metacarpals*	10
	Subtotal = 80	*Phalanges*	28
		Pelvic (hip) girdle	
		Hip, pelvic, or coxal bone	2
		Lower limbs (extremities)	
		Femur	2
		Patella	2
		Fibula	2
		Tibia	2
		Tarsals	14
		Metatarsals	10
		Phalanges	28
			Subtotal = 126
			Total = 206

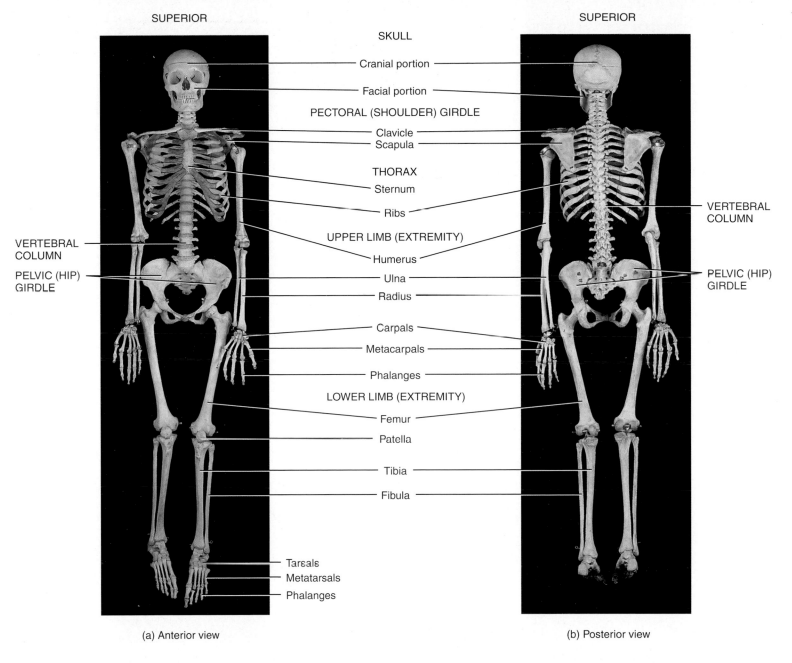

SUPERIOR

SUPERIOR

SKULL

Cranial portion

Facial portion

PECTORAL (SHOULDER) GIRDLE

Clavicle

Scapula

THORAX

Sternum

Ribs

UPPER LIMB (EXTREMITY)

Humerus

Ulna

Radius

Carpals

Metacarpals

Phalanges

LOWER LIMB (EXTREMITY)

Femur

Patella

Tibia

Fibula

VERTEBRAL COLUMN

PELVIC (HIP) GIRDLE

VERTEBRAL COLUMN

PELVIC (HIP) GIRDLE

Tarsals

Metatarsals

Phalanges

(a) Anterior view

(b) Posterior view

FIGURE 3.1
Complete skeleton

The Axial Skeleton

FIGURE 3.2
Skull

SUPERIOR

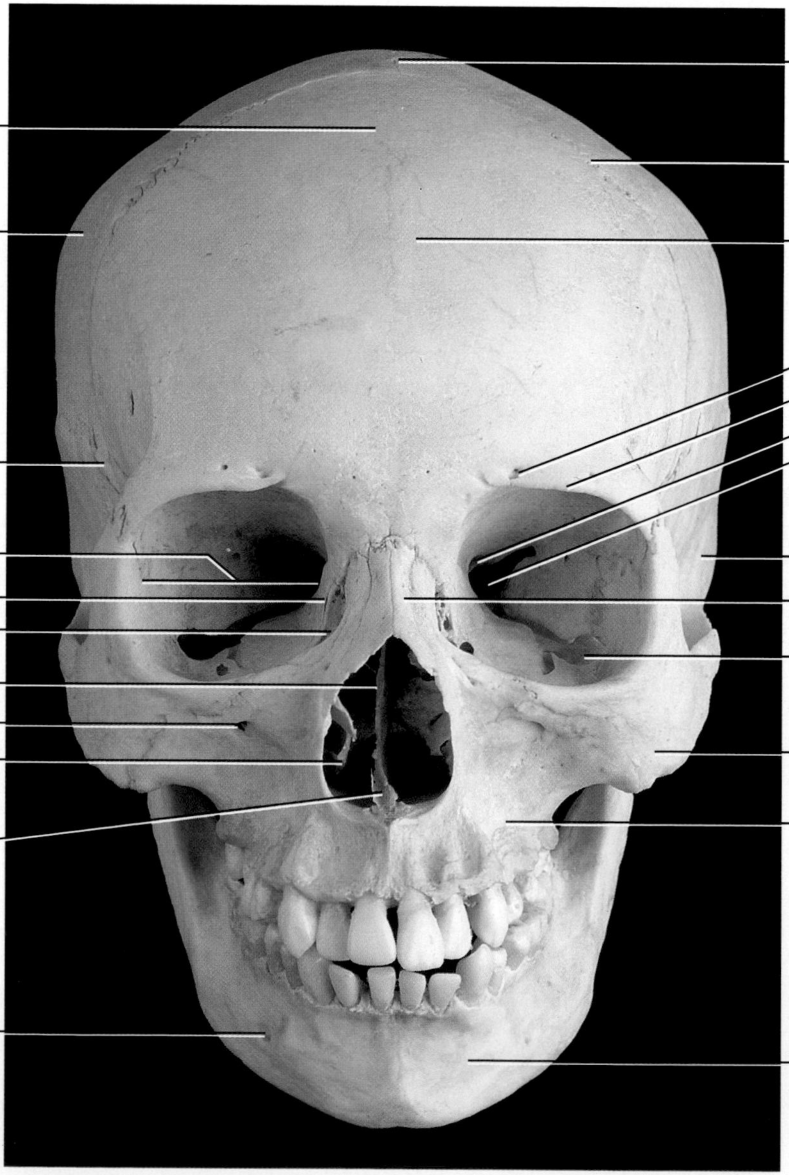

FRONTAL BONE
PARIETAL BONE
SPHENOID BONE
Orbit
ETHMOID BONE
LACRIMAL BONE
Perpendicular plate
Infraorbital foramen
INFERIOR NASAL CONCHA
VOMER
Mental foramen
Sagittal suture
Coronal suture
Frontal squama
Supraorbital foramen
Supraorbital margin
Optic foramen
Superior orbital fissure
TEMPORAL BONE
NASAL BONE
Inferior orbital fissure
ZYGOMATIC BONE
MAXILLA
MANDIBLE

INFERIOR

Anterior view

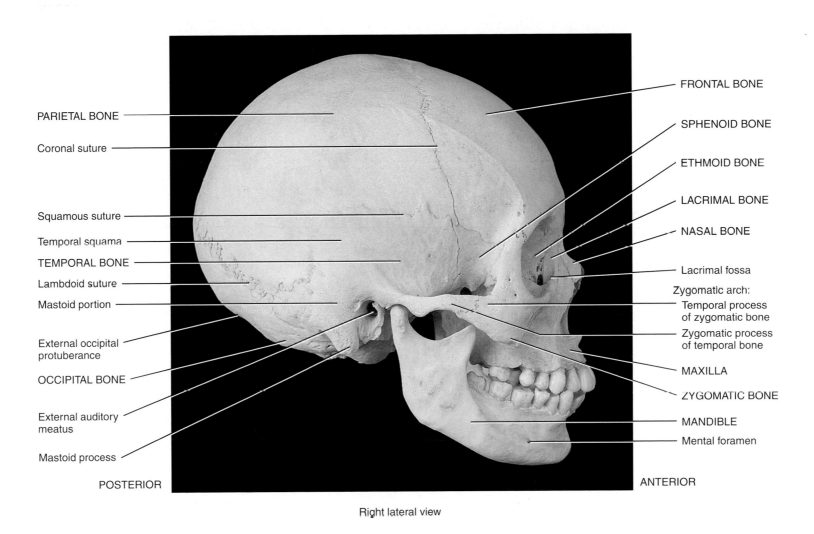

PARIETAL BONE

Coronal suture

Squamous suture

Temporal squama

TEMPORAL BONE

Lambdoid suture

Mastoid portion

External occipital
protuberance

OCCIPITAL BONE

External auditory
meatus

Mastoid process

FRONTAL BONE

SPHENOID BONE

ETHMOID BONE

LACRIMAL BONE

NASAL BONE

Lacrimal fossa

Zygomatic arch:

Temporal process
of zygomatic bone

Zygomatic process
of temporal bone

MAXILLA

ZYGOMATIC BONE

MANDIBLE

Mental foramen

POSTERIOR

ANTERIOR

Right lateral view

FIGURE 3.3
Skull

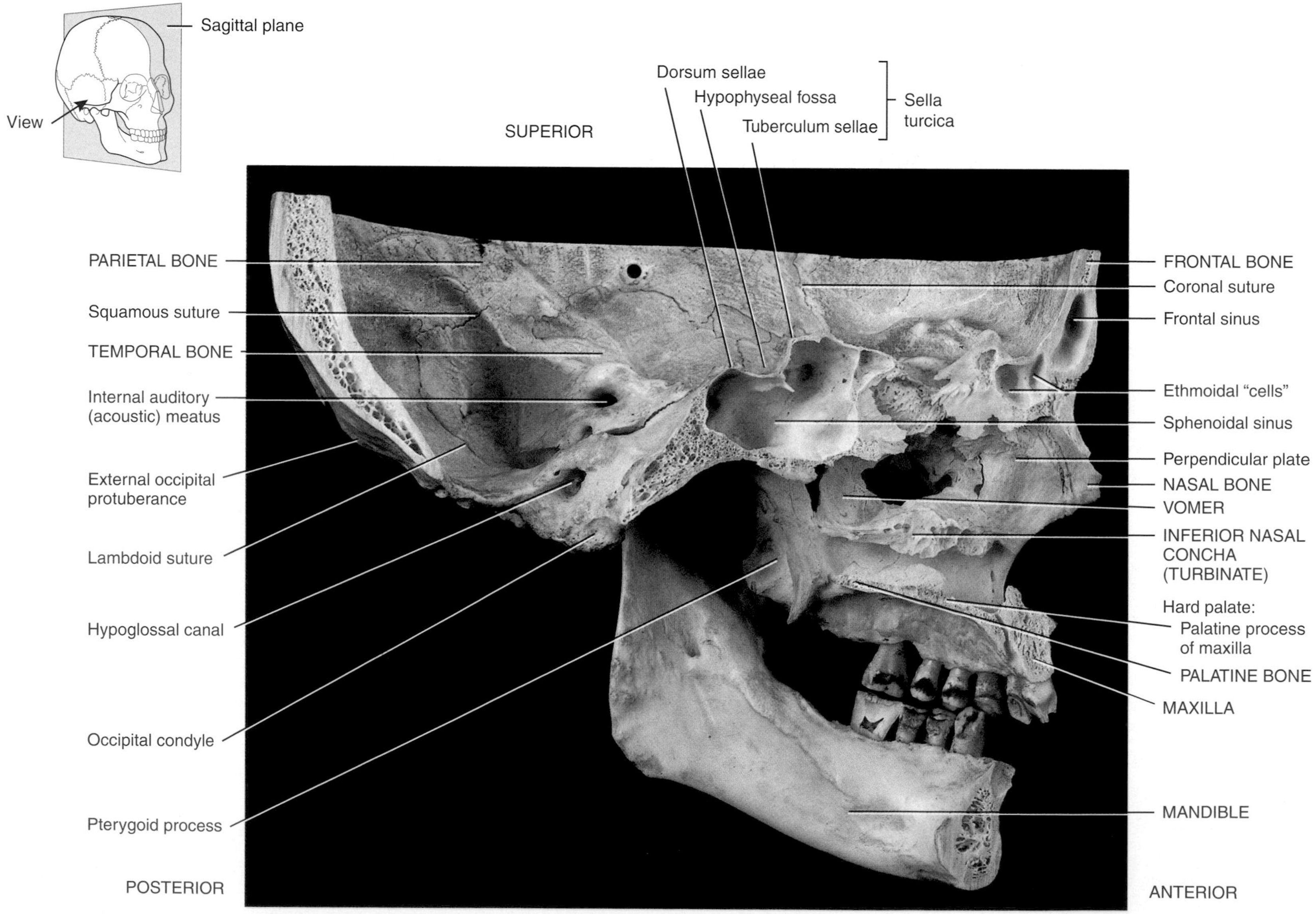

Sagittal plane

View

Dorsum sellae

Hypophyseal fossa — Sella turcica

Tuberculum sellae

SUPERIOR

PARIETAL BONE

Squamous suture

TEMPORAL BONE

Internal auditory (acoustic) meatus

External occipital protuberance

Lambdoid suture

Hypoglossal canal

Occipital condyle

Pterygoid process

POSTERIOR

FRONTAL BONE

Coronal suture

Frontal sinus

Ethmoidal "cells"

Sphenoidal sinus

Perpendicular plate

NASAL BONE

VOMER

INFERIOR NASAL CONCHA (TURBINATE)

Hard palate:
    Palatine process of maxilla

PALATINE BONE

MAXILLA

MANDIBLE

ANTERIOR

Medial view of sagittal section

FIGURE 3.4
Skull

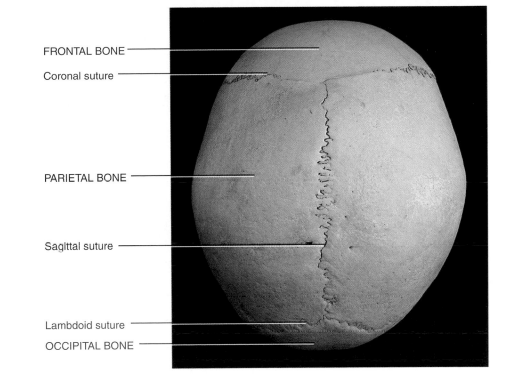

SUPERIOR

Sagittal suture

PARIETAL BONE

Lambdoid suture

OCCIPITAL BONE

External occipital
protuberance

INFERIOR

Posterior view

FIGURE 3.5
Skull

ANTERIOR

FRONTAL BONE

Coronal suture

PARIETAL BONE

Saglttal suture

Lambdoid suture

OCCIPITAL BONE

POSTERIOR

Supcrior vicw

FIGURE 3.6
Skull

Incisor teeth

MAXILLA:
Incisive foramen

Palatine process

Zygomatic arch

VOMER

SPHENOID BONE

Foramen ovale

Foramen spinosum

Mandibular (glenoid) fossa

Carotid foramen

Jugular foramen

Occipital condyle

Condylar canal

OCCIPITAL BONE

Inferior nuchal line

Superior nuchal line

View

ZYGOMATIC BONE

PALATINE BONE:
Horizontal plate
Greater palatine foramen
Lesser palatine foramen

Pterygoid processes

Articular tubercle

Foramen lacerum

Styloid process

Stylomastoid foramen

Mastoid process

Foramen magnum

Mastoid foramen

TEMPORAL BONE

Lambdoid suture

External occipital
protuberance

F I G U R E   3 . 7
Skull

POSTERIOR

Inferior view

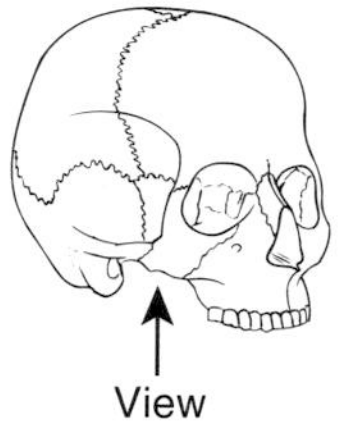

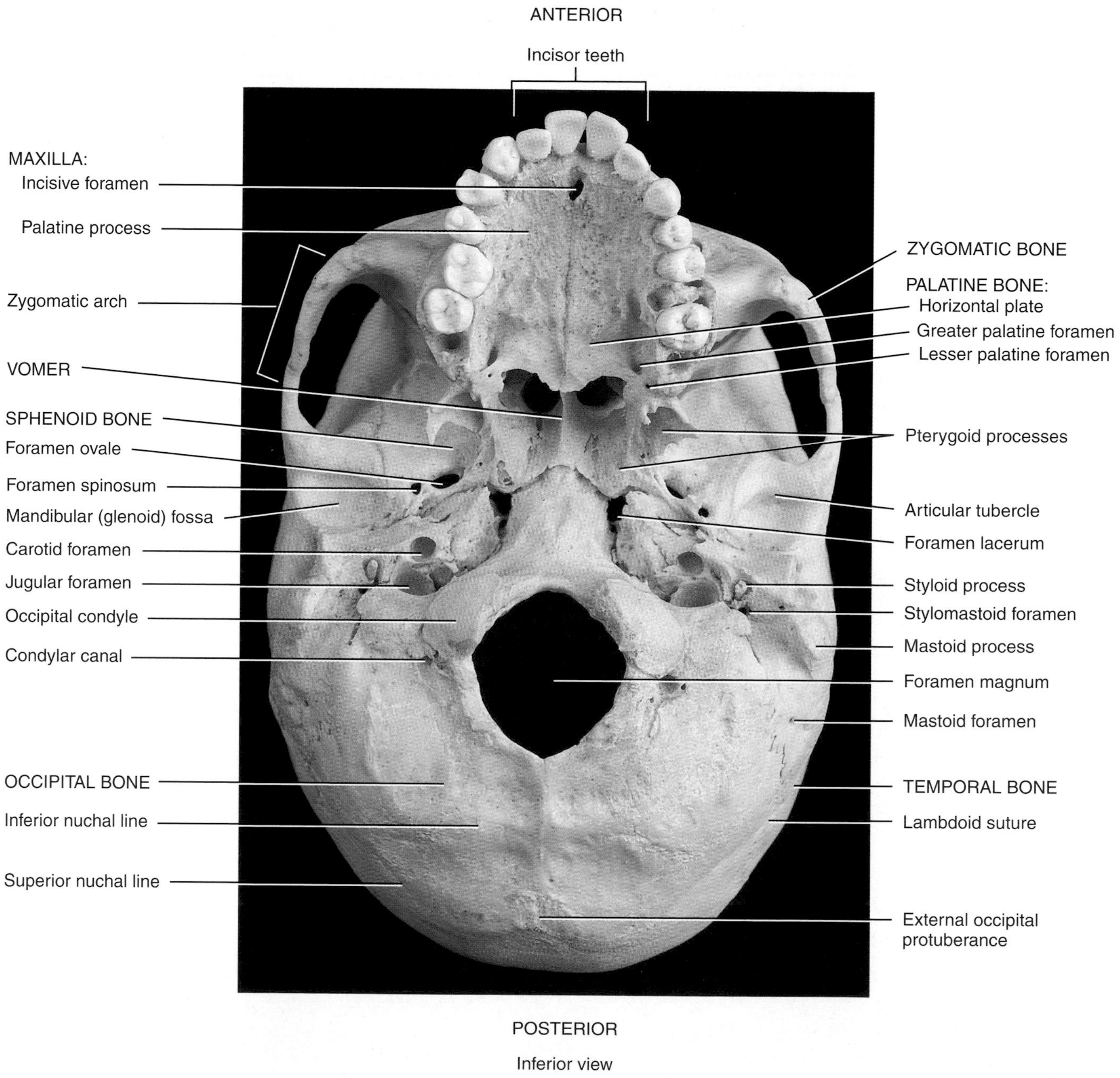

View

Transverse
plane

ANTERIOR

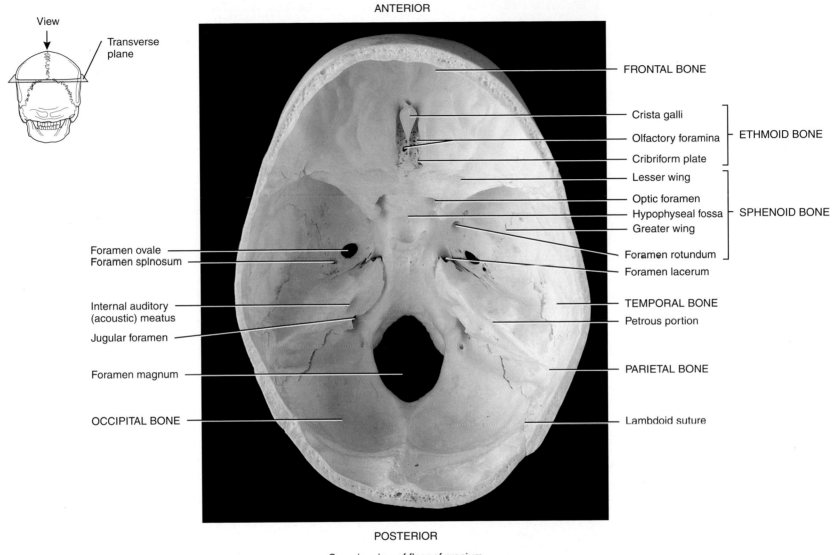

FRONTAL BONE

Crista galli
Olfactory foramina   ETHMOID BONE
Cribriform plate

Lesser wing
Optic foramen
Hypophyseal fossa   SPHENOID BONE
Greater wing
Foramen rotundum

Foramen ovale
Foramen splnosum

Foramen lacerum

Internal auditory
(acoustic) meatus

TEMPORAL BONE
Petrous portion

Jugular foramen

Foramen magnum

PARIETAL BONE

OCCIPITAL BONE

Lambdoid suture

POSTERIOR

Superior view of floor of cranium

F I G U R E   3 . 8
Skull

FIGURE 3.9 *Sphenoid bone*

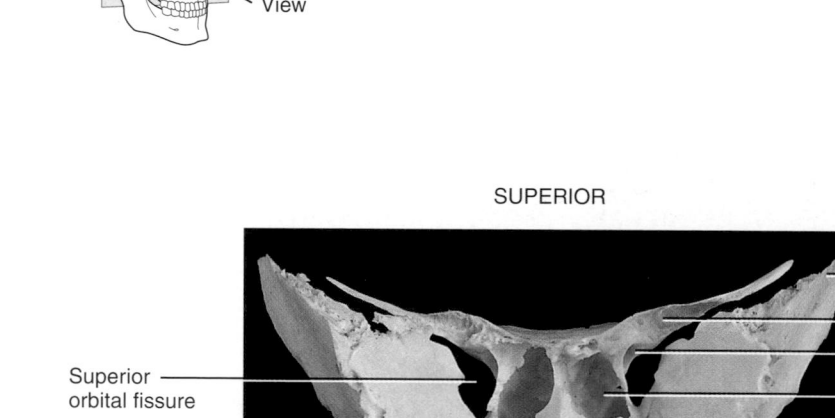

Frontal plane

View

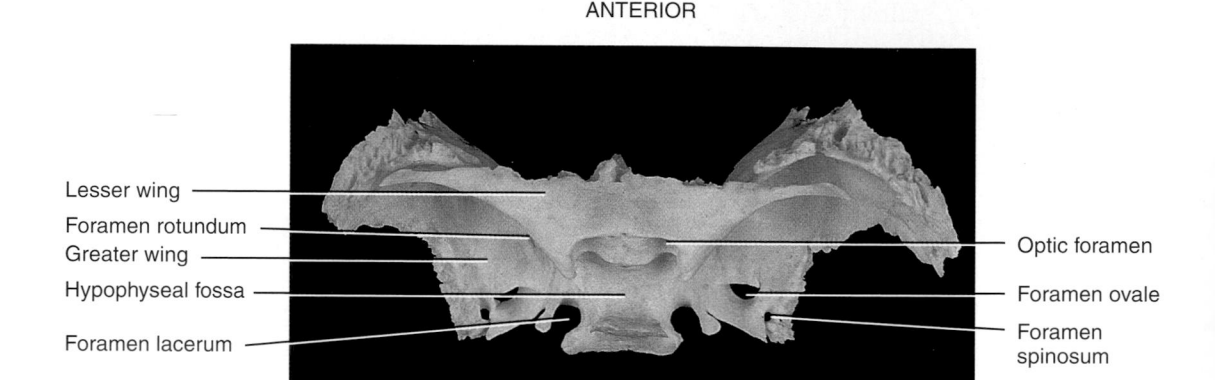

ANTERIOR

Lesser wing
Foramen rotundum
Greater wing
Hypophyseal fossa
Foramen lacerum

Optic foramen
Foramen ovale
Foramen spinosum

POSTERIOR

(a) Superior view

SUPERIOR

Superior orbital fissure

Foramen rotundum

Greater wing
Lesser wing
Optic foramen
Sphenoidal sinus
Body

Pterygoid processes

INFERIOR

(b) Anterior view

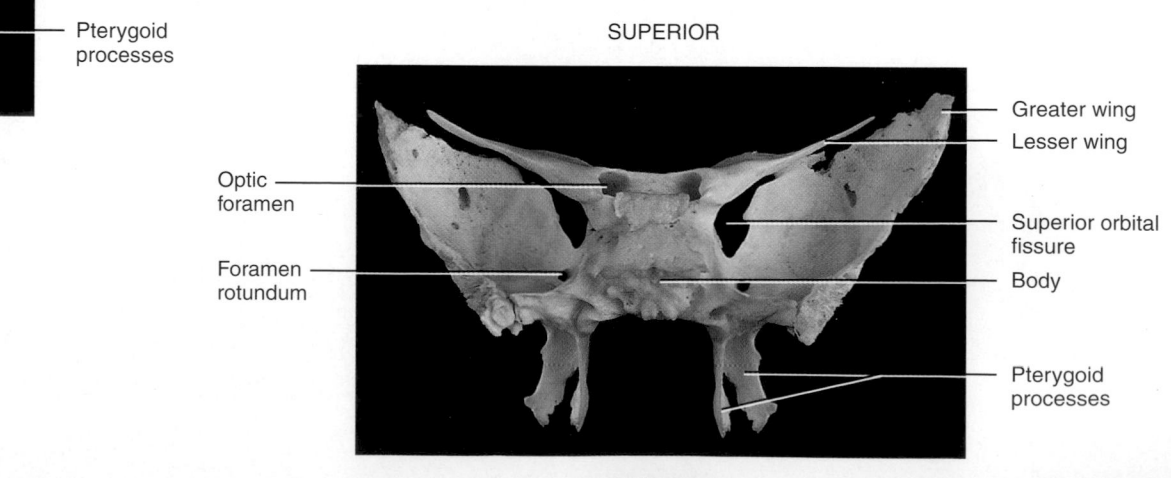

SUPERIOR

Optic foramen

Foramen rotundum

Greater wing
Lesser wing

Superior orbital fissure

Body

Pterygoid processes

INFERIOR

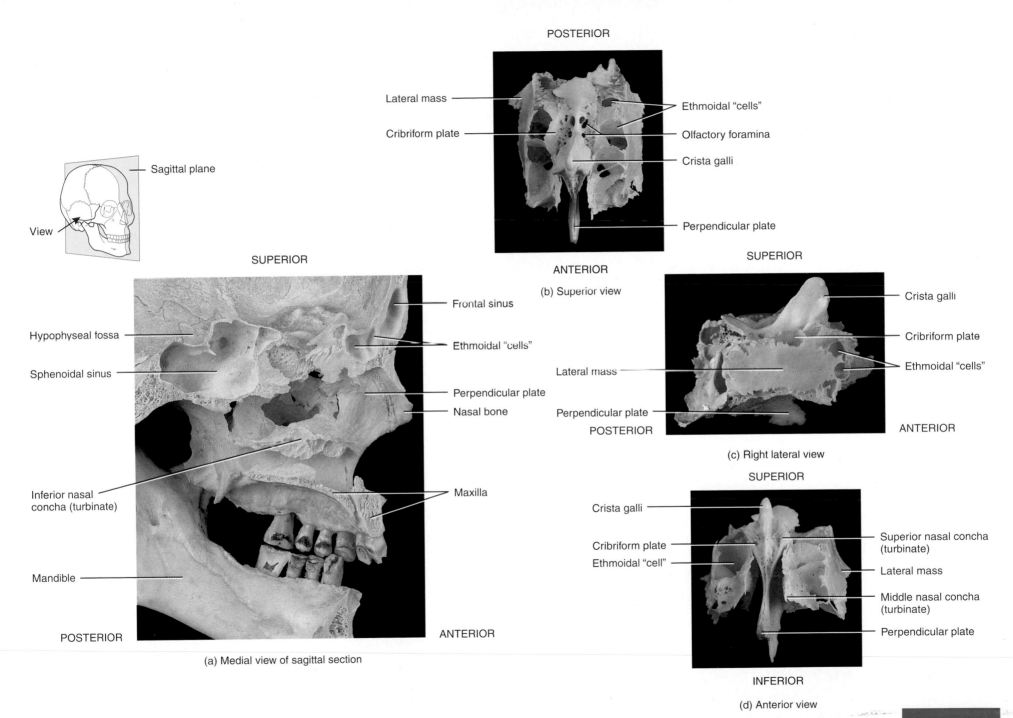

POSTERIOR

Lateral mass

Ethmoidal "cells"

Cribriform plate

Olfactory foramina

Crista galli

Perpendicular plate

ANTERIOR

(b) Superior view

Sagittal plane

View

SUPERIOR

Hypophyseal fossa

Frontal sinus

Ethmoidal "cells"

Sphenoidal sinus

Perpendicular plate

Nasal bone

Inferior nasal concha (turbinate)

Maxilla

Mandible

POSTERIOR

ANTERIOR

(a) Medial view of sagittal section

SUPERIOR

Crista galli

Cribriform plate

Lateral mass

Ethmoidal "cells"

Perpendicular plate

POSTERIOR

ANTERIOR

(c) Right lateral view

SUPERIOR

Crista galli

Cribriform plate

Ethmoidal "cell"

Superior nasal concha (turbinate)

Lateral mass

Middle nasal concha (turbinate)

Perpendicular plate

INFERIOR

(d) Anterior view

FIGURE 3.10 *Ethmoid bone*

FIGURE 3.11
Right orbit

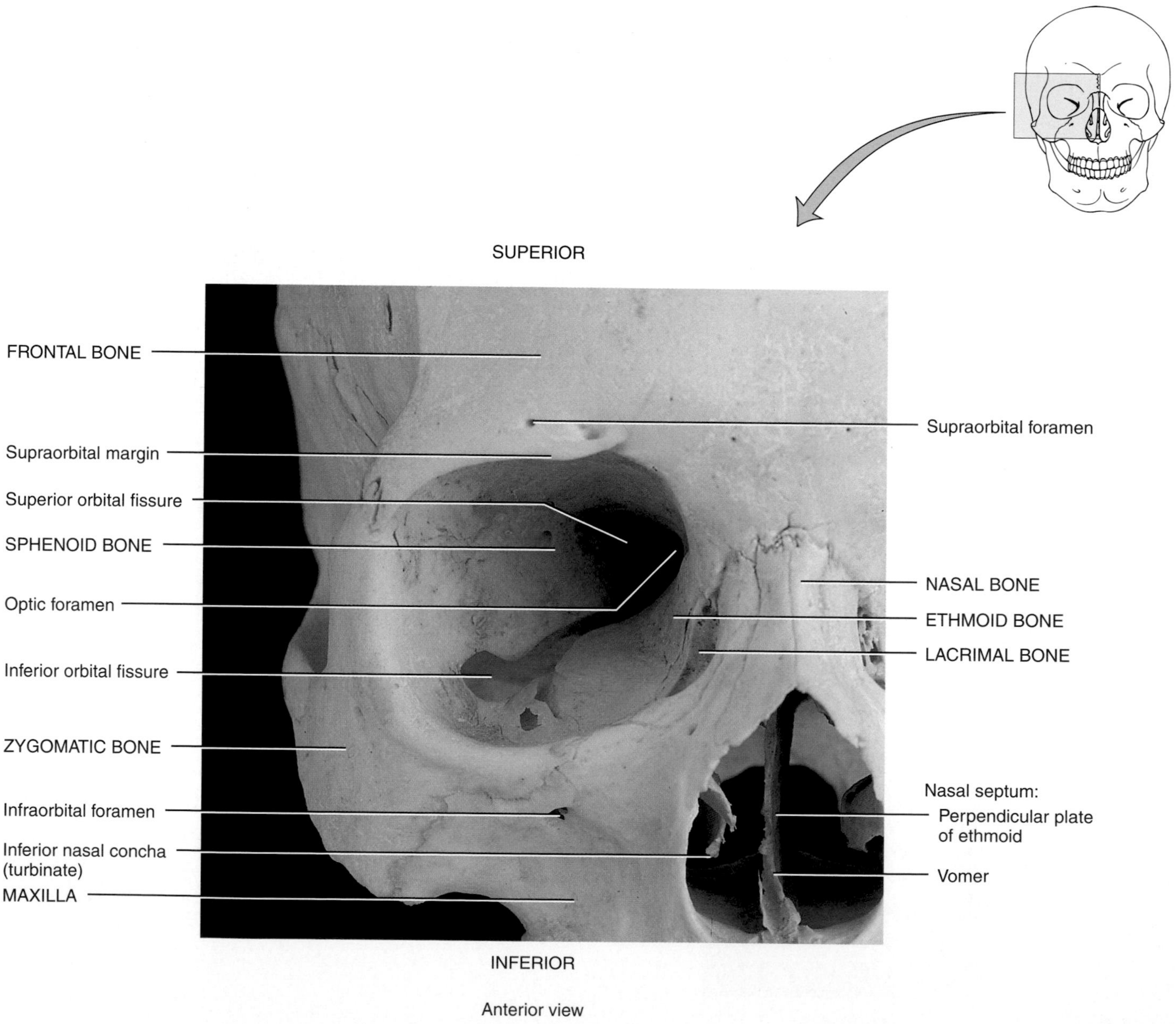

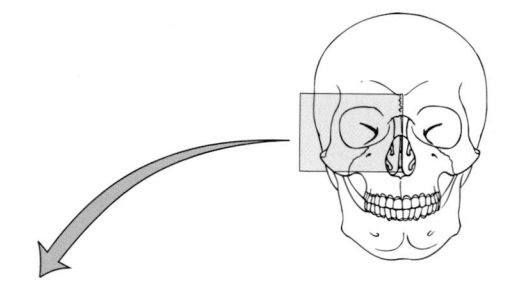

SUPERIOR

FRONTAL BONE

Supraorbital foramen

Supraorbital margin

Superior orbital fissure

SPHENOID BONE

NASAL BONE

Optic foramen

ETHMOID BONE

LACRIMAL BONE

Inferior orbital fissure

ZYGOMATIC BONE

Nasal septum:

Infraorbital foramen

Perpendicular plate
of ethmoid

Inferior nasal concha
(turbinate)

Vomer

MAXILLA

INFERIOR

Anterior view

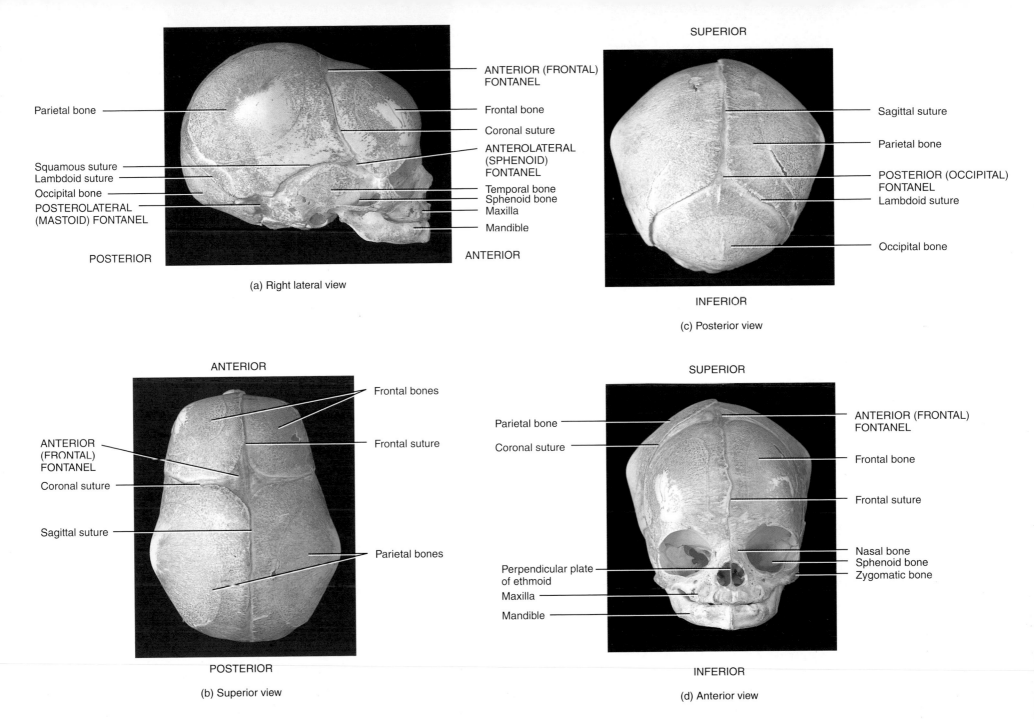

Parietal bone

Squamous suture
Lambdoid suture
Occipital bone
POSTEROLATERAL
(MASTOID) FONTANEL

POSTERIOR

ANTERIOR (FRONTAL)
FONTANEL
Frontal bone
Coronal suture
ANTEROLATERAL
(SPHENOID)
FONTANEL
Temporal bone
Sphenoid bone
Maxilla
Mandible

ANTERIOR

(a) Right lateral view

SUPERIOR

Sagittal suture
Parietal bone
POSTERIOR (OCCIPITAL)
FONTANEL
Lambdoid suture

Occipital bone

INFERIOR

(c) Posterior view

ANTERIOR

ANTERIOR
(FRONTAL)
FONTANEL
Coronal suture

Sagittal suture

Frontal bones

Frontal suture

Parietal bones

POSTERIOR

(b) Superior view

SUPERIOR

Parietal bone
Coronal suture

ANTERIOR (FRONTAL)
FONTANEL
Frontal bone

Frontal suture

Perpendicular plate
of ethmoid
Maxilla
Mandible

Nasal bone
Sphenoid bone
Zygomatic bone

INFERIOR

(d) Anterior view

FIGURE 3.12
Fontanels of a fetal skull

FIGURE 3.14
Auditory ossicles

INFERIOR

MEDIAL

LATERAL

Stapes

Malleus

Incus

SUPERIOR

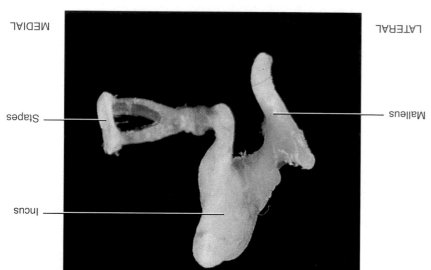

FIGURE 3.13
Hyoid bone

Superior view

ANTERIOR

Body

Lesser horn

Greater horn

POSTERIOR

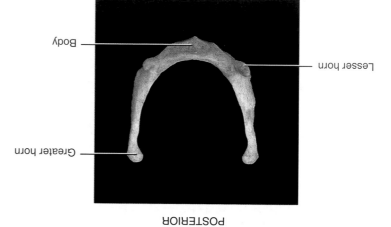

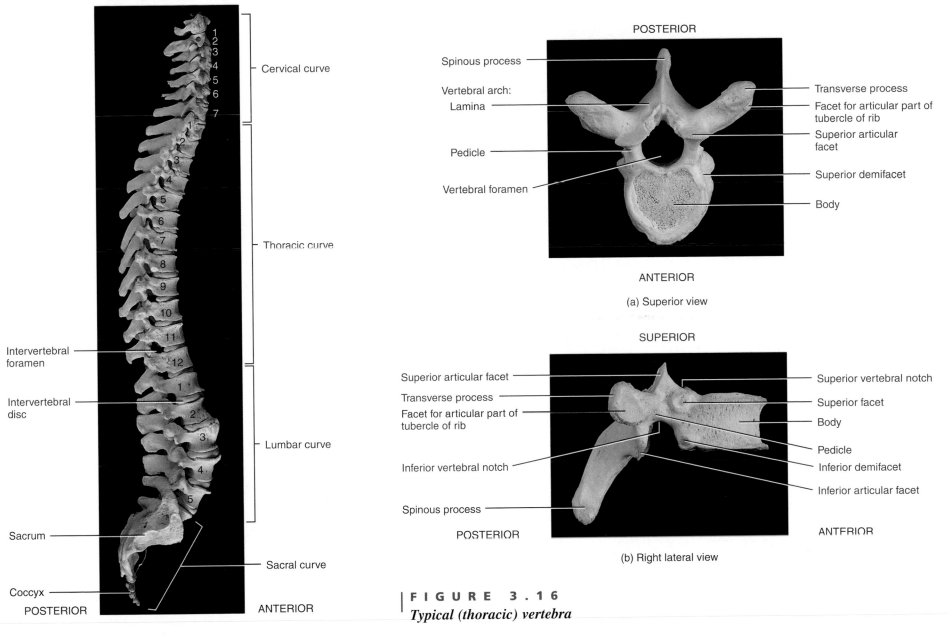

SUPERIOR

1
2
3
4
5
6
7

Cervical curve

1
2
3
4
5
6
7
8
9
10
11
12

Thoracic curve

Intervertebral foramen

Intervertebral disc

1
2
3
4
5

Lumbar curve

Sacrum

Sacral curve

Coccyx

POSTERIOR

ANTERIOR

FIGURE 3.15
Vertebral column

POSTERIOR

Spinous process

Vertebral arch:
Lamina

Transverse process

Facet for articular part of tubercle of rib

Superior articular facet

Pedicle

Superior demifacet

Vertebral foramen

Body

ANTERIOR

(a) Superior view

SUPERIOR

Superior articular facet

Superior vertebral notch

Transverse process

Superior facet

Facet for articular part of tubercle of rib

Body

Pedicle

Inferior vertebral notch

Inferior demifacet

Inferior articular facet

Spinous process

POSTERIOR

ANTERIOR

(b) Right lateral view

FIGURE 3.16
Typical (thoracic) vertebra

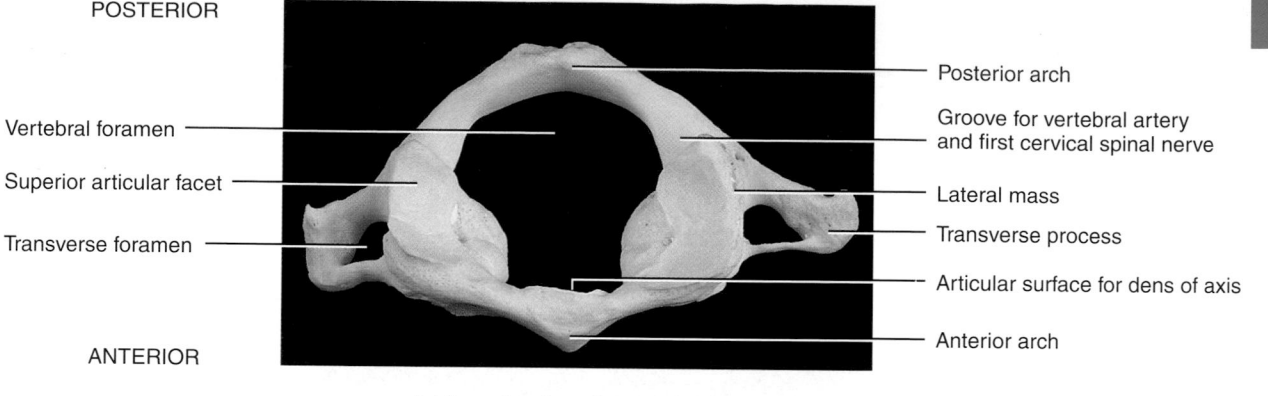

POSTERIOR

Vertebral foramen — Posterior arch

Superior articular facet — Groove for vertebral artery and first cervical spinal nerve

Transverse foramen — Lateral mass

— Transverse process

— Articular surface for dens of axis

ANTERIOR — Anterior arch

(a) Superior view of the atlas (C1)

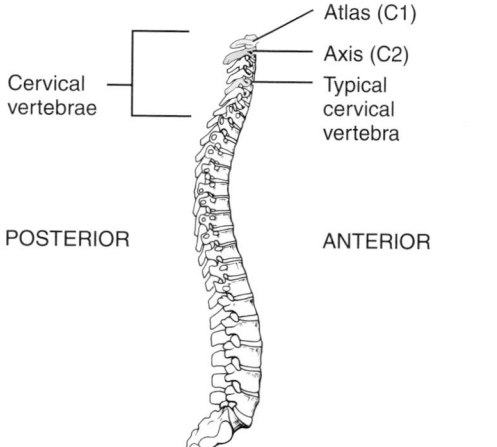

Atlas (C1)

Axis (C2)

Typical cervical vertebra

Cervical vertebrae

POSTERIOR   ANTERIOR

Location of cervical vertebrae

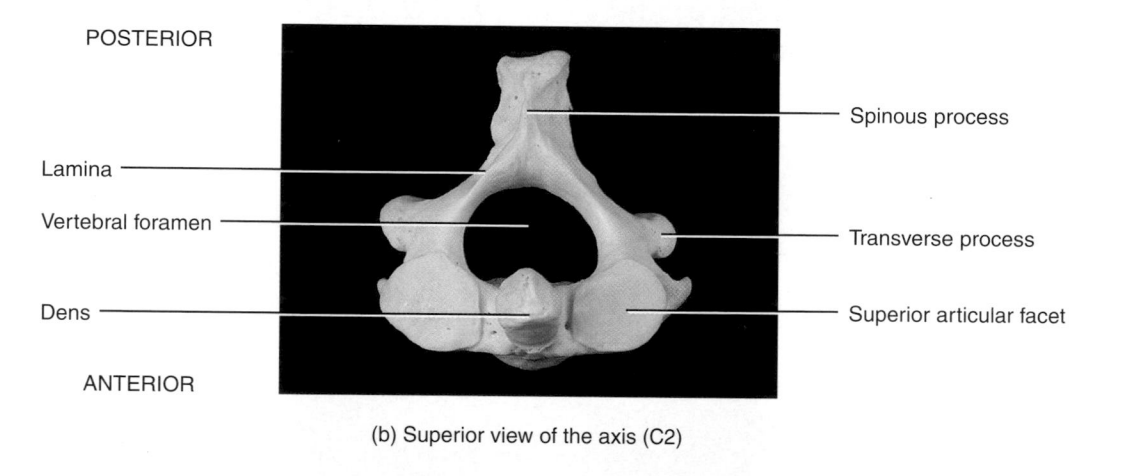

POSTERIOR

Spinous process

Lamina

Vertebral foramen — Transverse process

Dens — Superior articular facet

ANTERIOR

(b) Superior view of the axis (C2)

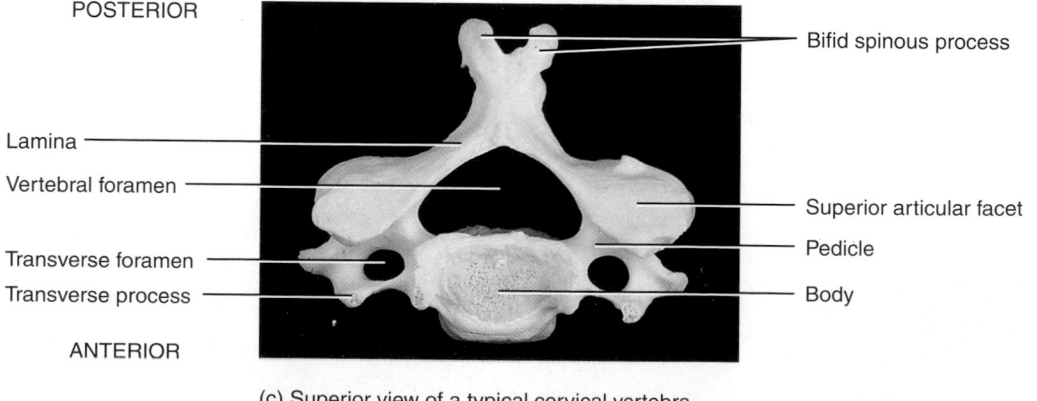

POSTERIOR

Bifid spinous process

Lamina

Vertebral foramen — Superior articular facet

Transverse foramen — Pedicle

Transverse process — Body

ANTERIOR

(c) Superior view of a typical cervical vertebra

FIGURE 3.17
Cervical vertebrae

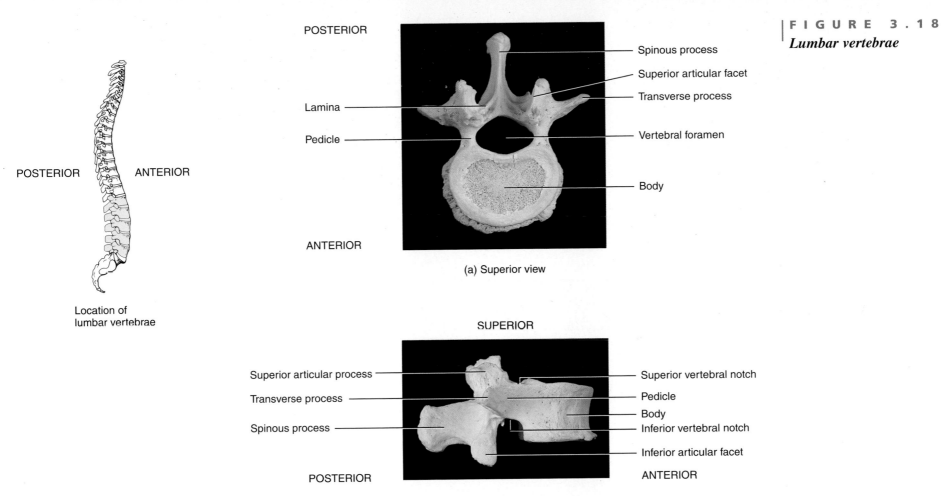

POSTERIOR

Spinous process

Superior articular facet

Transverse process

Lamina

Pedicle

Vertebral foramen

Body

ANTERIOR

(a) Superior view

POSTERIOR   ANTERIOR

Location of
lumbar vertebrae

SUPERIOR

Superior articular process

Superior vertebral notch

Transverse process

Pedicle

Body

Spinous process

Inferior vertebral notch

Inferior articular facet

POSTERIOR   ANTERIOR

(b) Right lateral view

FIGURE 3.18
Lumbar vertebrae

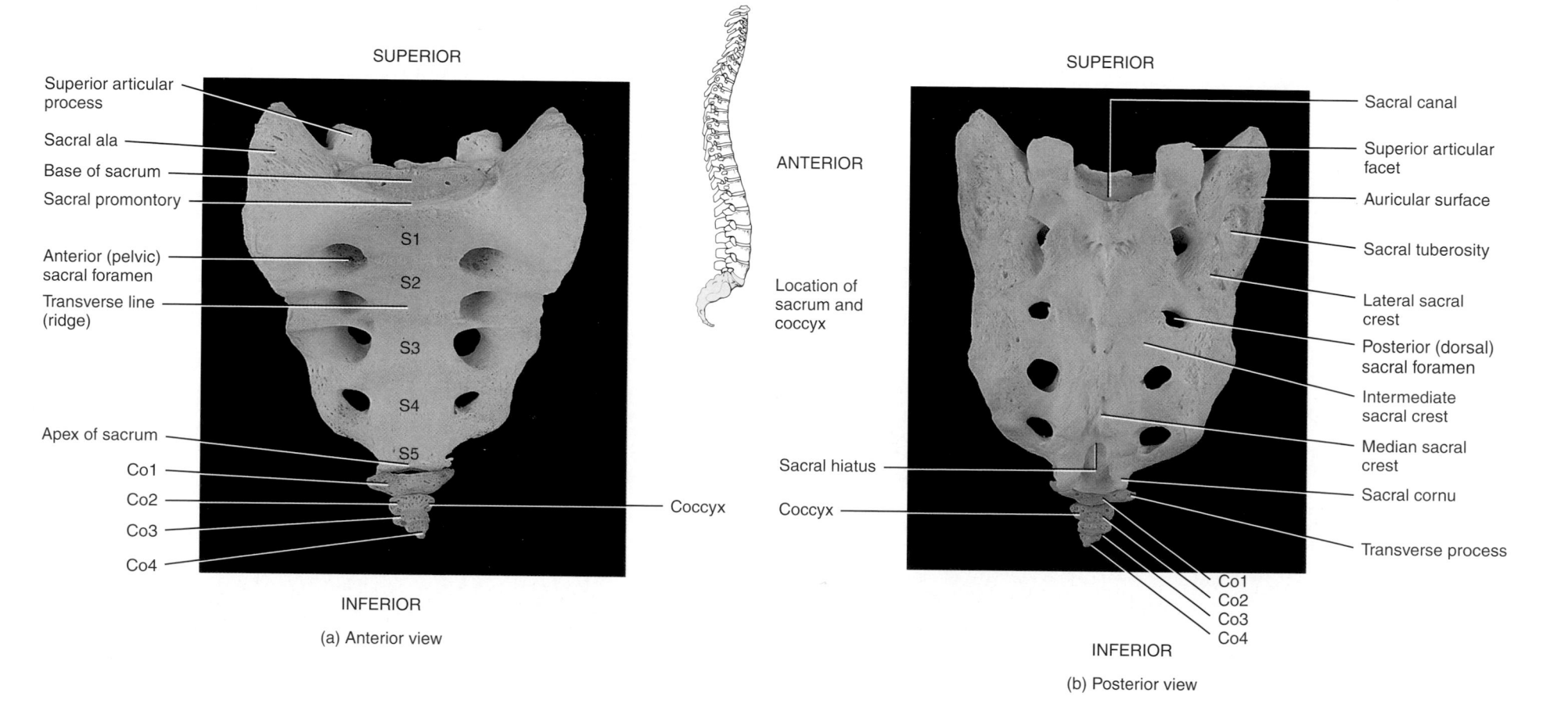

SUPERIOR

Superior articular process

Sacral ala

Base of sacrum

Sacral promontory

S1

S2

Anterior (pelvic) sacral foramen

Transverse line (ridge)

S3

S4

Apex of sacrum

Co1

S5

Co2

Co3

Co4

Coccyx

INFERIOR

(a) Anterior view

ANTERIOR

Location of sacrum and coccyx

SUPERIOR

Sacral canal

Superior articular facet

Auricular surface

Sacral tuberosity

Lateral sacral crest

Posterior (dorsal) sacral foramen

Intermediate sacral crest

Median sacral crest

Sacral cornu

Transverse process

Sacral hiatus

Coccyx

Co1
Co2
Co3
Co4

INFERIOR

(b) Posterior view

FIGURE 3.19
Sacrum and coccyx

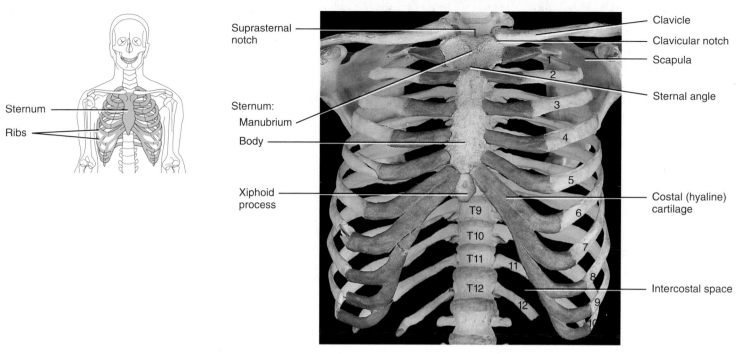

Sternum

Ribs

SUPERIOR

Suprasternal notch

Sternum:

Manubrium

Body

Xiphoid process

T9

T10

T11

T12

1
2
3
4
5
6
7
11
8
12
9
10

Clavicle

Clavicular notch

Scapula

Sternal angle

Costal (hyaline) cartilage

Intercostal space

INFERIOR

Anterior view

FIGURE 3.20
Anterior view of skeleton of thorax

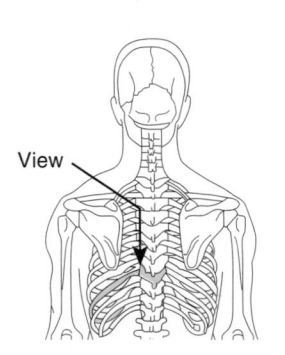

View

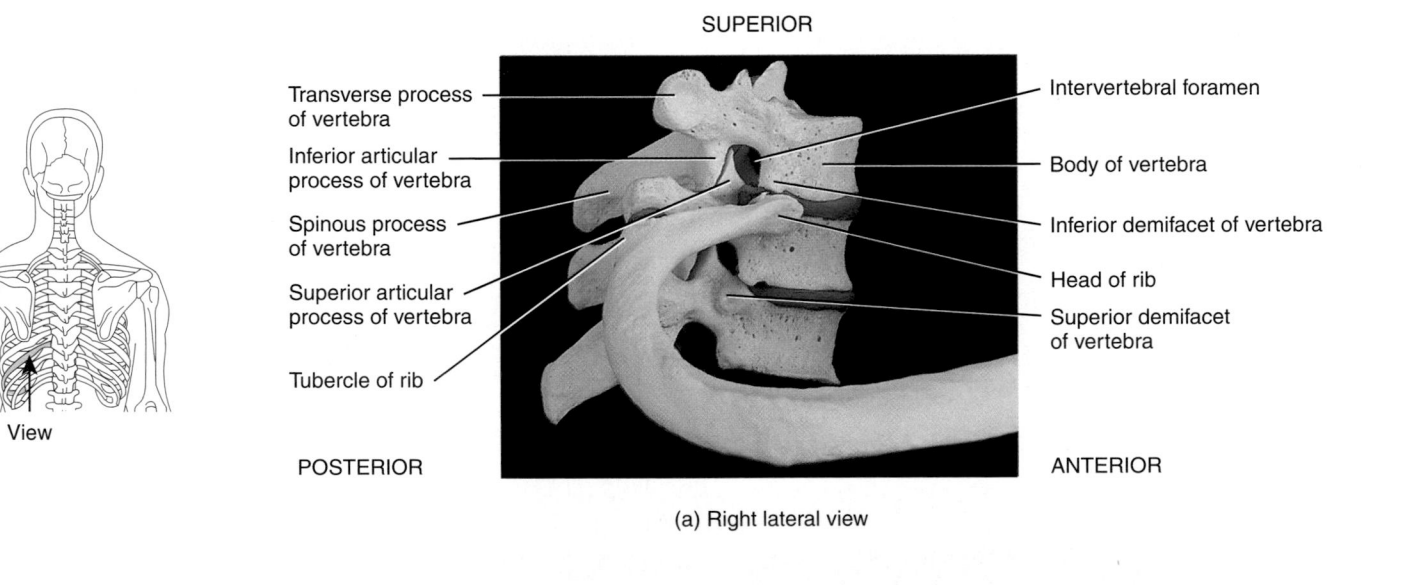

SUPERIOR

Transverse process of vertebra

Inferior articular process of vertebra

Spinous process of vertebra

Superior articular process of vertebra

Tubercle of rib

Intervertebral foramen

Body of vertebra

Inferior demifacet of vertebra

Head of rib

Superior demifacet of vertebra

POSTERIOR

ANTERIOR

(a) Right lateral view

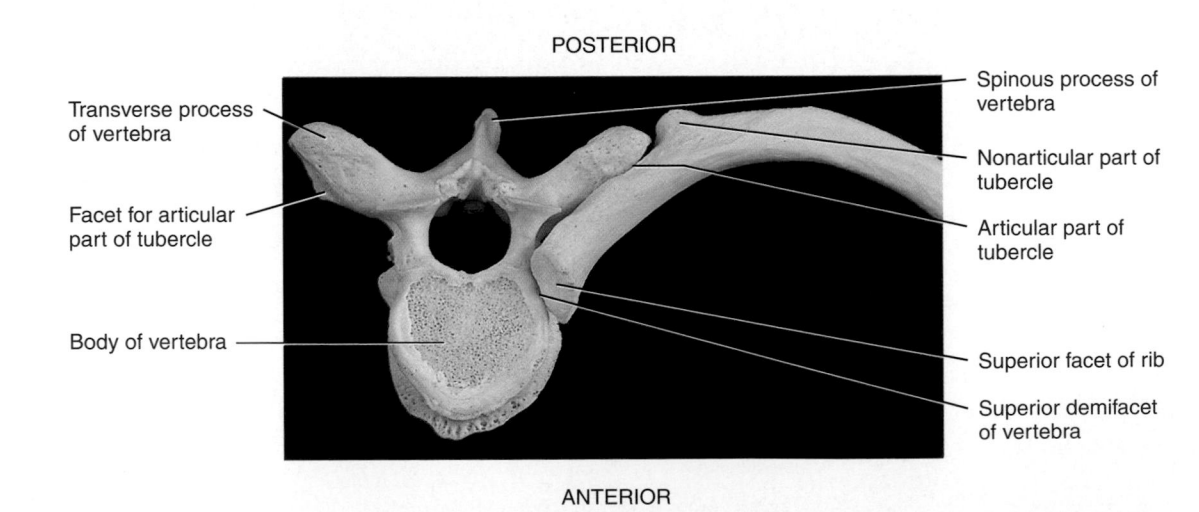

POSTERIOR

Transverse process of vertebra

Facet for articular part of tubercle

Body of vertebra

Spinous process of vertebra

Nonarticular part of tubercle

Articular part of tubercle

Superior facet of rib

Superior demifacet of vertebra

ANTERIOR

(b) Superior view

FIGURE 3.21
Articulation of a thoracic vertebra with a rib

The Appendicular Skeleton

FIGURE 3.22

Right scapula

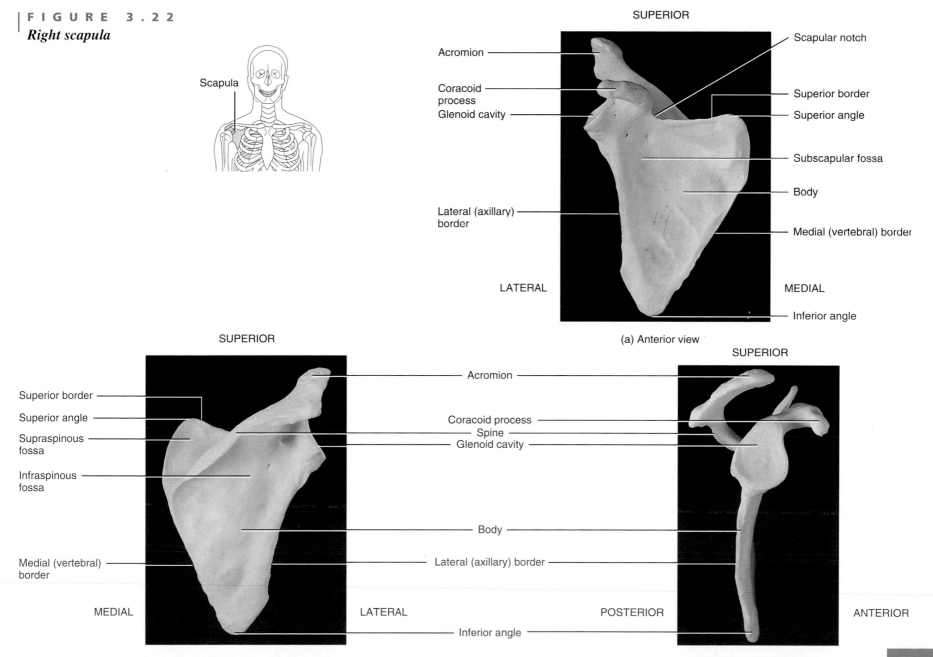

Scapula

SUPERIOR

Acromion

Coracoid process

Glenoid cavity

Scapular notch

Superior border

Superior angle

Subscapular fossa

Body

Lateral (axillary) border

Medial (vertebral) border

LATERAL

MEDIAL

Inferior angle

(a) Anterior view

SUPERIOR

Superior border

Superior angle

Supraspinous fossa

Infraspinous fossa

Medial (vertebral) border

Acromion

Coracoid process

Spine

Glenoid cavity

Body

Lateral (axillary) border

MEDIAL

LATERAL

Inferior angle

(b) Posterior view

SUPERIOR

POSTERIOR

ANTERIOR

(c) Lateral border view

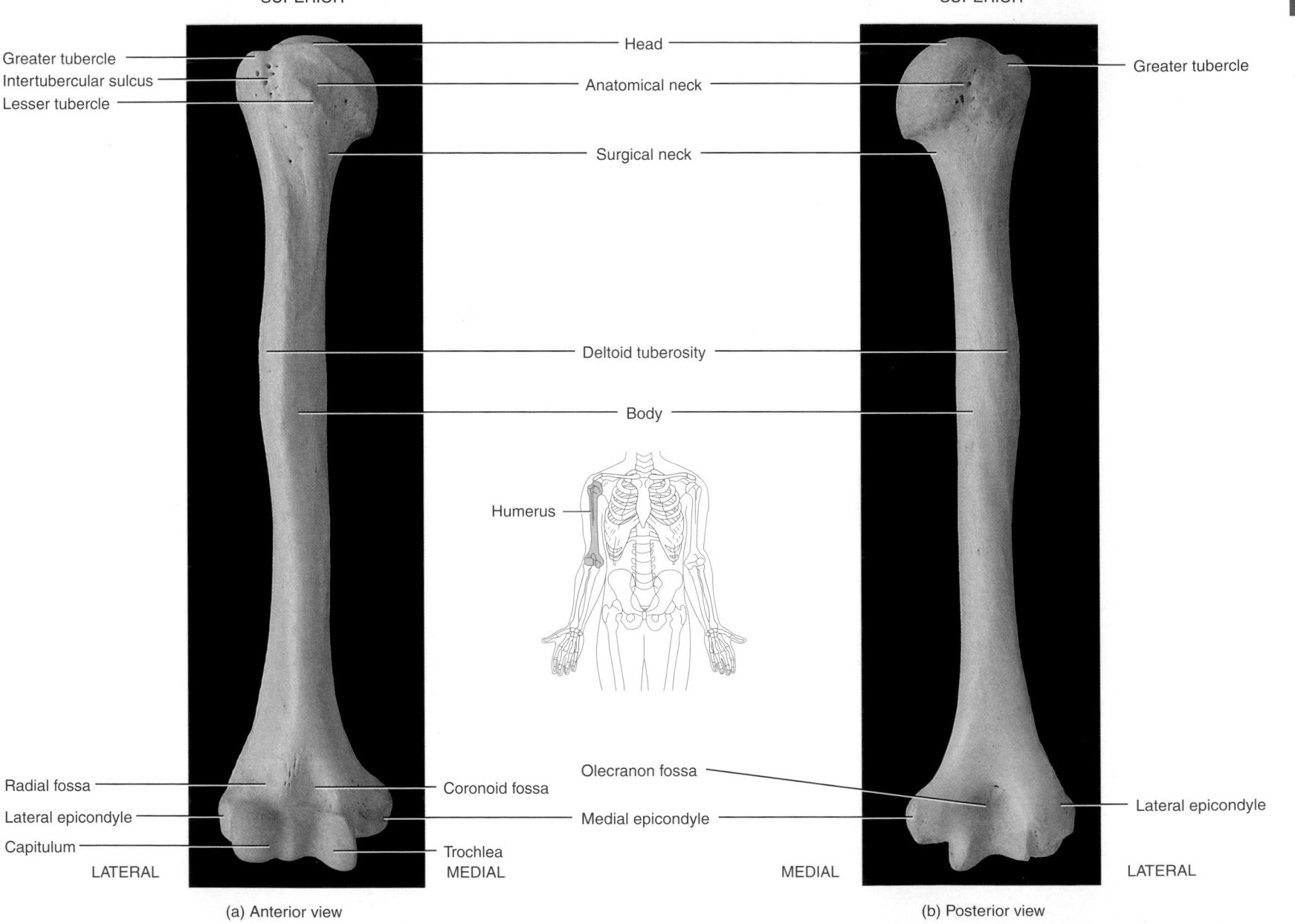

SUPERIOR

Greater tubercle
Intertubercular sulcus
Lesser tubercle

Head
Anatomical neck

Surgical neck

Deltoid tuberosity

Body

Humerus

Radial fossa
Lateral epicondyle
Capitulum

LATERAL

Coronoid fossa

Medial epicondyle

Trochlea
MEDIAL

(a) Anterior view

SUPERIOR

Greater tubercle

Olecranon fossa

Lateral epicondyle

MEDIAL

LATERAL

(b) Posterior view

FIGURE 3.23
Right humerus

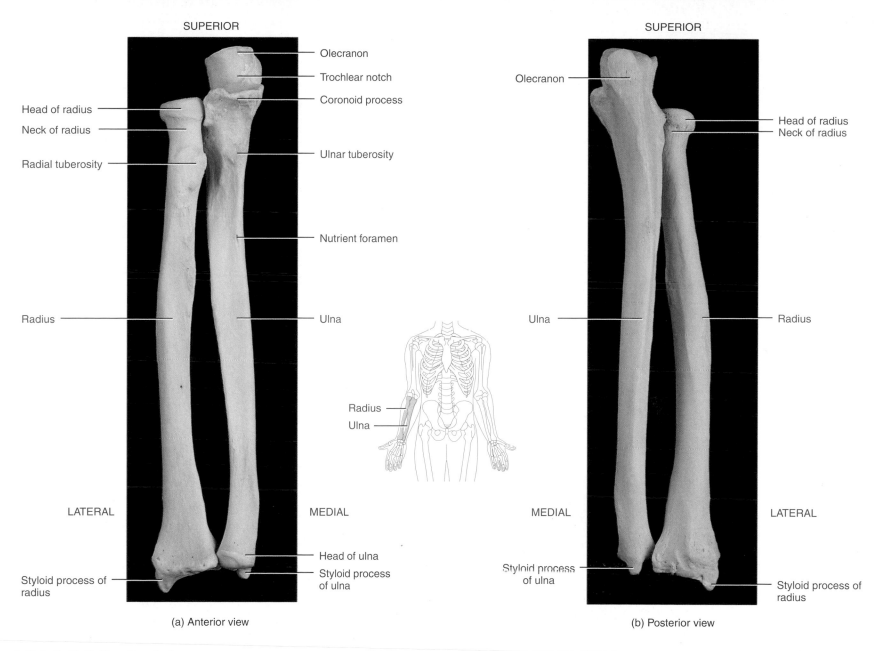

SUPERIOR

Olecranon

Trochlear notch

Coronoid process

Head of radius

Neck of radius

Radial tuberosity

Ulnar tuberosity

Nutrient foramen

Radius

Ulna

Radius

Ulna

LATERAL

MEDIAL

Head of ulna

Styloid process of radius

Styloid process of ulna

(a) Anterior view

SUPERIOR

Olecranon

Head of radius

Neck of radius

Ulna

Radius

MEDIAL

LATERAL

Styloid process of ulna

Styloid process of radius

(b) Posterior view

FIGURE   3.24

Right ulna and radius

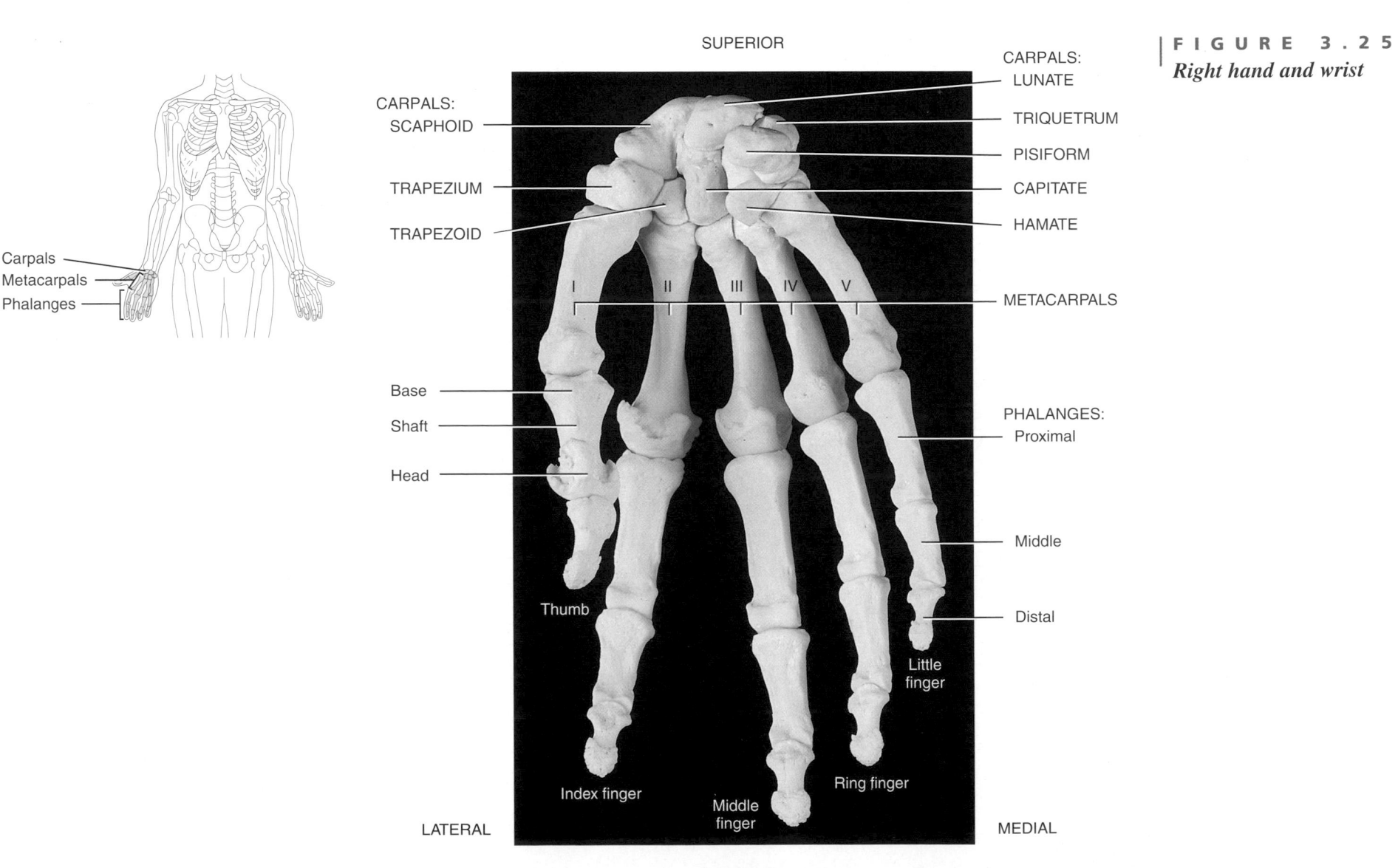

FIGURE 3.25
Right hand and wrist

SUPERIOR

Carpals
Metacarpals
Phalanges

CARPALS:
SCAPHOID

CARPALS:
LUNATE

TRIQUETRUM

PISIFORM

TRAPEZIUM

CAPITATE

TRAPEZOID

HAMATE

I II III IV V

METACARPALS

Base

Shaft

Head

PHALANGES:
Proximal

Middle

Distal

Thumb

Little
finger

Index finger

Middle
finger

Ring finger

LATERAL

MEDIAL

Anterior view

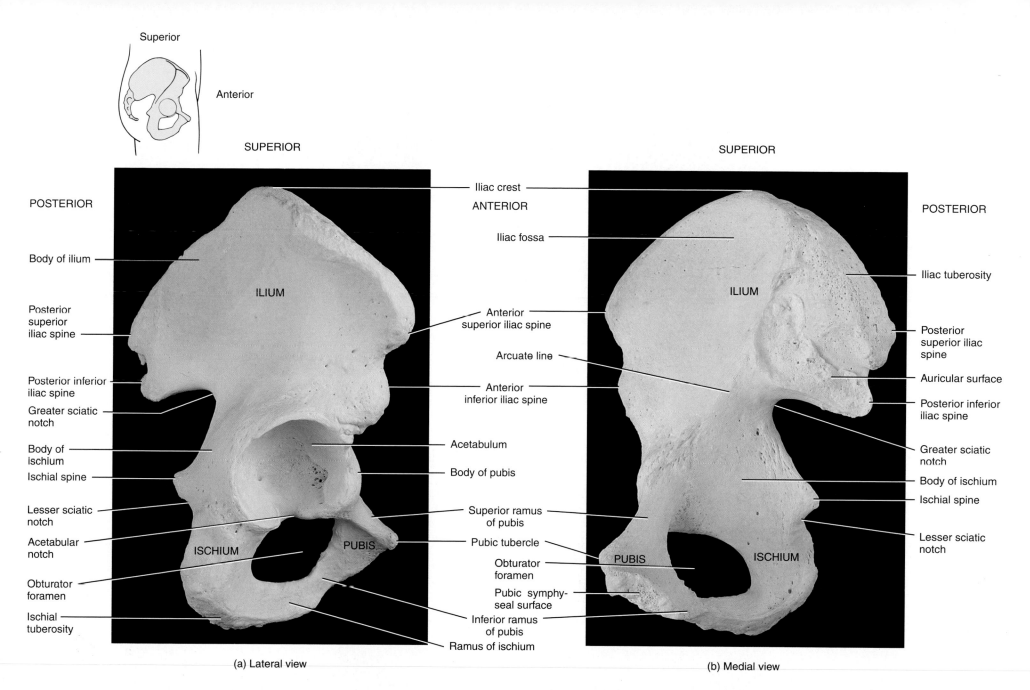

Superior

Anterior

SUPERIOR

Iliac crest

ANTERIOR

POSTERIOR

POSTERIOR

Iliac fossa

Body of ilium

ILIUM

ILIUM

Iliac tuberosity

Posterior superior iliac spine

Anterior superior iliac spine

Posterior superior iliac spine

Posterior inferior iliac spine

Arcuate line

Auricular surface

Greater sciatic notch

Anterior inferior iliac spine

Posterior inferior iliac spine

Body of Ischium

Acetabulum

Greater sciatic notch

Ischial spine

Body of pubis

Body of ischium

Lesser sciatic notch

Superior ramus of pubis

Ischial spine

Acetabular notch

ISCHIUM

PUBIS

Pubic tubercle

PUBIS

ISCHIUM

Lesser sciatic notch

Obturator foramen

Obturator foramen

Pubic symphyseal surface

Ischial tuberosity

Inferior ramus of pubis

Ramus of ischium

(a) Lateral view

(b) Medial view

FIGURE 3.26
Right hip bone

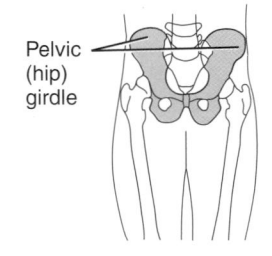

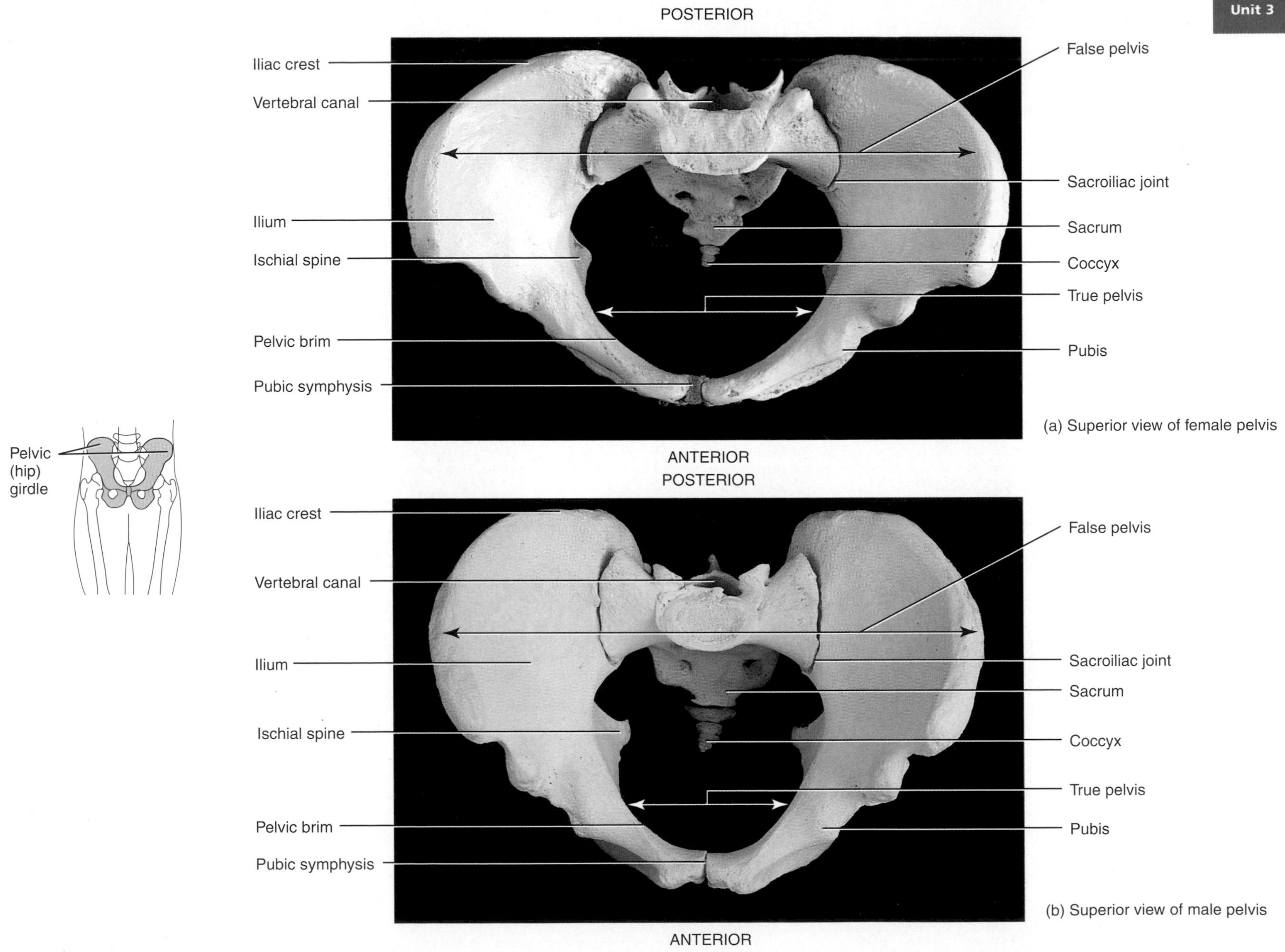

Pelvic (hip) girdle

POSTERIOR

Iliac crest

Vertebral canal

Ilium

Ischial spine

Pelvic brim

Pubic symphysis

False pelvis

Sacroiliac joint

Sacrum

Coccyx

True pelvis

Pubis

(a) Superior view of female pelvis

ANTERIOR

POSTERIOR

Iliac crest

Vertebral canal

Ilium

Ischial spine

Pelvic brim

Pubic symphysis

False pelvis

Sacroiliac joint

Sacrum

Coccyx

True pelvis

Pubis

(b) Superior view of male pelvis

ANTERIOR

FIGURE 3.27
Pelvis

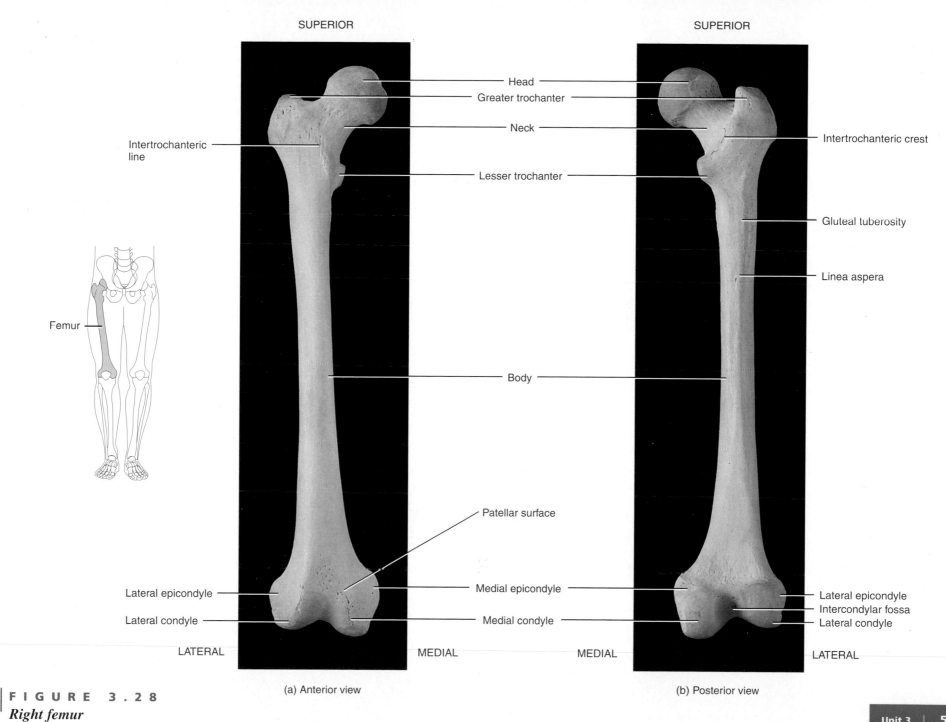

SUPERIOR

SUPERIOR

Head

Greater trochanter

Neck

Intertrochanteric line

Intertrochanteric crest

Lesser trochanter

Gluteal tuberosity

Linea aspera

Femur

Body

Patellar surface

Lateral epicondyle

Medial epicondyle

Lateral epicondyle
Intercondylar fossa
Lateral condyle

Lateral condyle

Medial condyle

LATERAL

MEDIAL

MEDIAL

LATERAL

(a) Anterior view

(b) Posterior view

F I G U R E   3 . 2 8
Right femur

SUPERIOR

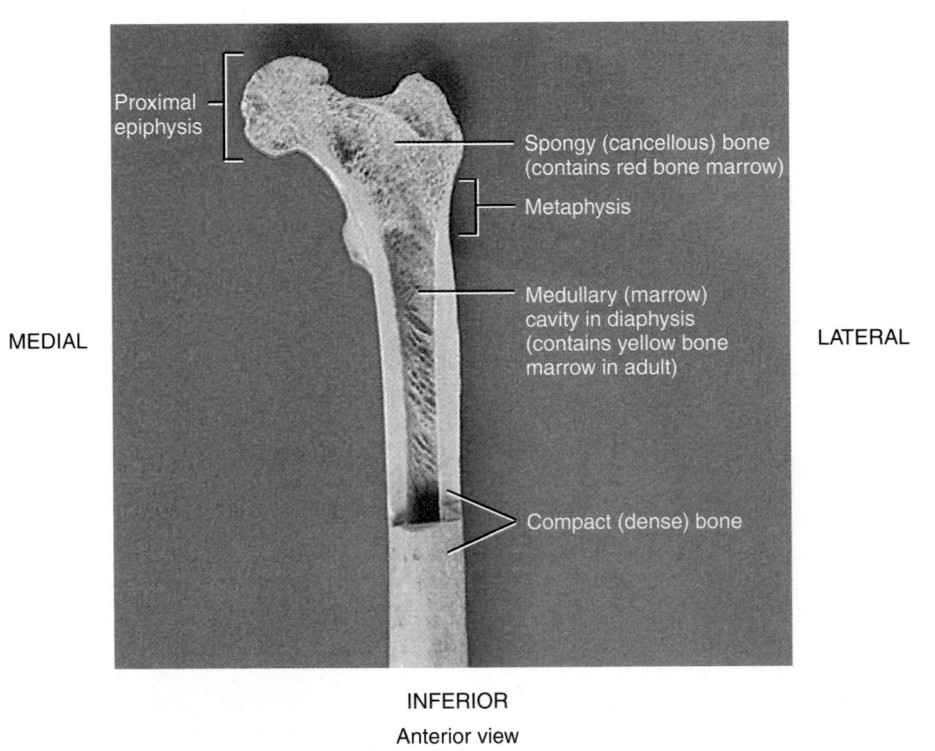

Proximal epiphysis

Spongy (cancellous) bone (contains red bone marrow)

Metaphysis

Medullary (marrow) cavity in diaphysis (contains yellow bone marrow in adult)

MEDIAL

LATERAL

Compact (dense) bone

INFERIOR

Anterior view

FIGURE 3.29
Portion of a partially sectioned femur

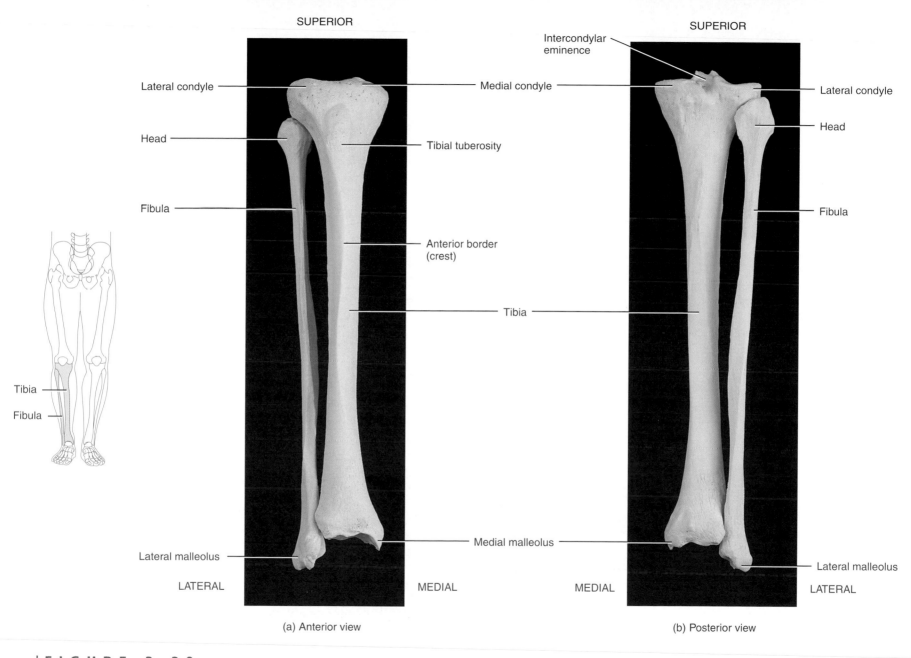

SUPERIOR

Lateral condyle

Head

Fibula

Lateral malleolus

LATERAL

Medial condyle

Tibial tuberosity

Anterior border
(crest)

Tibia

Medial malleolus

MEDIAL

(a) Anterior view

SUPERIOR

Intercondylar
eminence

Lateral condyle

Head

Fibula

Lateral malleolus

LATERAL

Medial malleolus

MEDIAL

(b) Posterior view

Tibia

Fibula

FIGURE 3.30
Right tibia and fibula

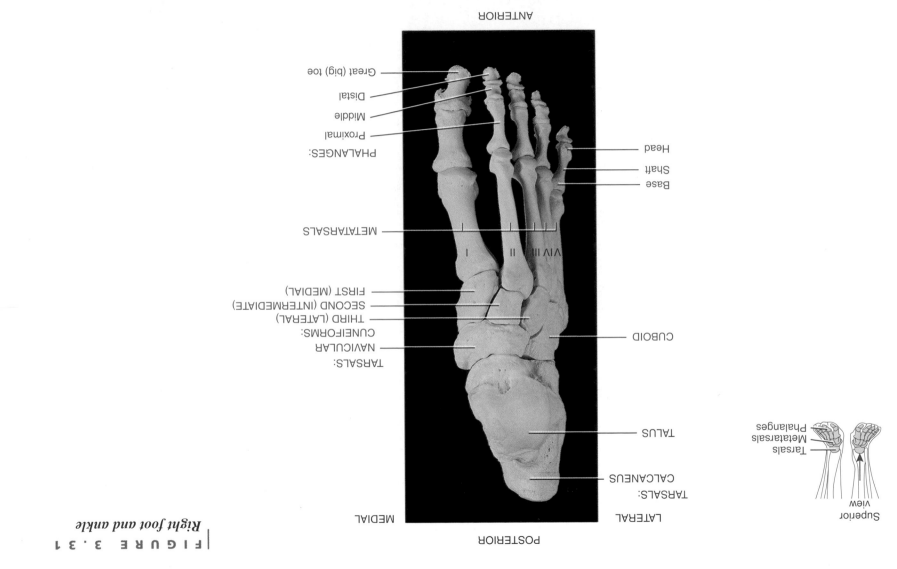

Superior view

ANTERIOR

Great (big) toe

Distal

Middle

Proximal

PHALANGES:

Head

Shaft

Base

METATARSALS

I   II   III   IV   V

FIRST (MEDIAL)

SECOND (INTERMEDIATE)

THIRD (LATERAL)

CUNEIFORMS:

NAVICULAR

TARSALS:

CUBOID

TALUS

CALCANEUS

TARSALS:

MEDIAL

LATERAL

POSTERIOR

Tarsals
Metatarsals
Phalanges

Superior
view

FIGURE 3.31
Right foot and ankle

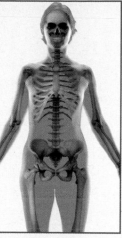

TABLE 4.1 | *Summary of Structural Categories and Functional Characteristics of Joints*

Structural Category	Description	Functional Characteristics	Example
FIBROUS JOINTS: Articulating bones held together by fibrous connective tissue; no synovial cavity.			
Suture	Articulating bones united by a thin layer of dense fibrous connective tissue; found between bones of the skull. With age, some sutures are replaced by a synostosis, in which the bones fuse across the former suture.	Synarthrosis (immovable).	Frontal suture.
Syndesmosis	Articulating bones are united by a dense fibrous connective tissue, either a ligament or an interosseous membrane.	Amphiarthrosis (slightly movable).	Distal tibiofibular joint.
Gomphosis	Articulating bones united by a periodontal ligament; cone-shaped peg fits into a socket.	Synarthrosis.	Roots of teeth in sockets of maxillae and mandible.
CARTILAGINOUS JOINTS: Articulating bones united by cartilage; no synovial cavity.			
Synchondrosis	Connecting material is hyaline cartilage; becomes a synostosis when bone elongation ceases.	Synarthrosis.	Epiphyseal plate at joint between the diaphysis and epiphysis of a long bone.
Symphysis	Connecting material is a broad, flat disc of fibrocartilage.	Amphiarthrosis.	Intervertebral joints and pubic symphysis.
SYNOVIAL JOINTS: Characterized by a synovial cavity, articular cartilage, and an articular capsule; may contain accessory ligaments, articular discs, and bursae.			
Planar	Articulating surfaces are flat or slightly curved.	Nonaxial diarthrosis (freely movable); gliding motion.	Intercarpal, intertarsal, sternocostal (between sternum and the 2nd–7th pairs of ribs), and vertebrocostal joints.
Hinge	Convex surface fits into a concave surface.	Monaxial diarthrosis; angular motion.	Elbow, ankle, and interphalangeal joints.
Pivot	Rounded or pointed surface fits into a ring formed partly by bone and partly by a ligament.	Monaxial diarthrosis; rotation.	Atlanto-axial and proximal radioulnar joints.
Condyloid	Oval-shaped projection fits into an oval-shaped depression.	Biaxial diarthrosis; angular motion.	Radiocarpal and metacarpophalangeal joints.
Saddle	Articular surface of one bone is saddle-shaped, and the articular surface of the other bone "sits" in the saddle.	Biaxial diarthrosis; angular motion.	Carpometacarpal joint between trapezium and thumb.
Ball-and-socket	Ball-like surface fits into a cuplike depression.	Multiaxial diarthrosis; angular motion rotation.	Shoulder and hip joints.

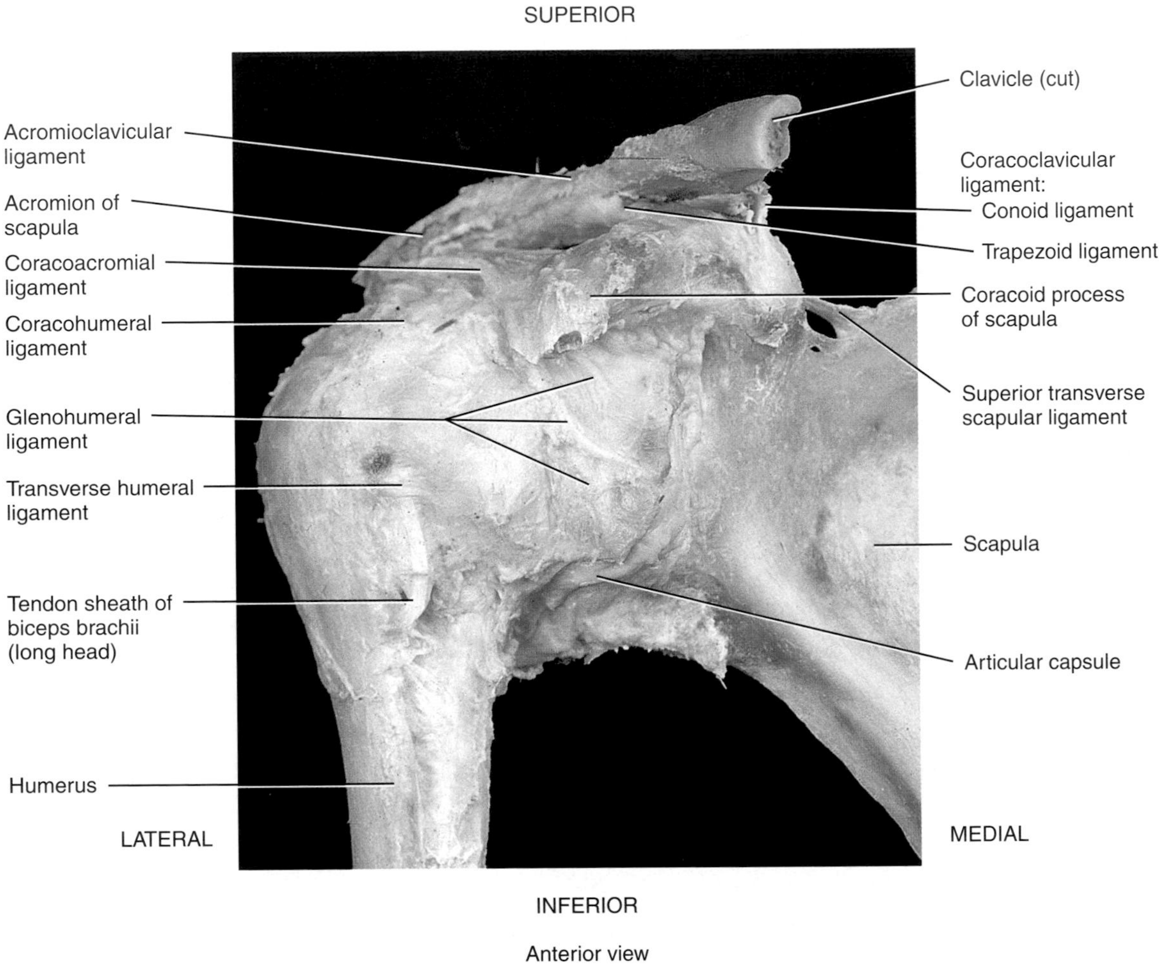

FIGURE 4.1
Right shoulder (glenohumeral) joint

SUPERIOR

Clavicle (cut)

Acromioclavicular
ligament

Coracoclavicular
ligament:
    Conoid ligament

Acromion of
scapula

Coracoacromial
ligament

Trapezoid ligament

Coracohumeral
ligament

Coracoid process
of scapula

Glenohumeral
ligament

Superior transverse
scapular ligament

Transverse humeral
ligament

Scapula

Tendon sheath of
biceps brachii
(long head)

Articular capsule

Humerus

LATERAL

MEDIAL

INFERIOR

Anterior view

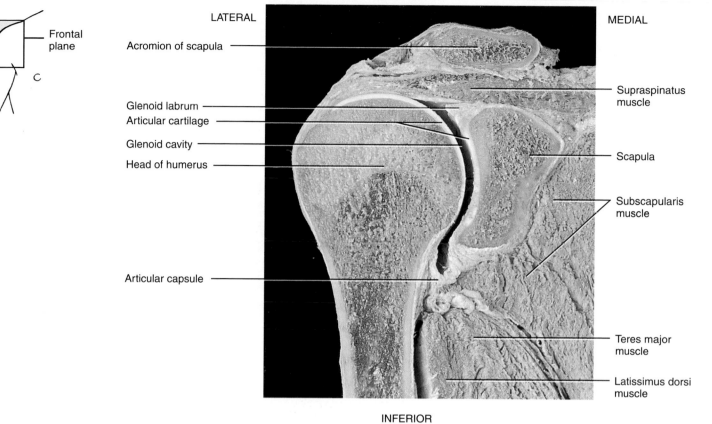

SUPERIOR

LATERAL   MEDIAL

Frontal
plane

C

Acromion of scapula

Supraspinatus
muscle

Glenoid labrum
Articular cartilage
Glenoid cavity

Scapula

Head of humerus

Subscapularis
muscle

Articular capsule

Teres major
muscle

Latissimus dorsi
muscle

INFERIOR

Frontal section

FIGURE 4.2
Right shoulder (glenohumeral) joint

FIGURE 4.3
Right elbow joint

SUPERIOR

Humerus

Capitulum (hidden under joint capsule)

Lateral epicondyle

Articular capsule

Medial epicondyle

Trochlea (hidden under joint capsule)

Radial collateral ligament

Annular ligament

Ulnar collateral ligament

Tendon of biceps brachii muscle (cut)

Radial tuberosity

Radius

Ulna

Interosseous membrane

LATERAL

MEDIAL

INFERIOR

Anterior view

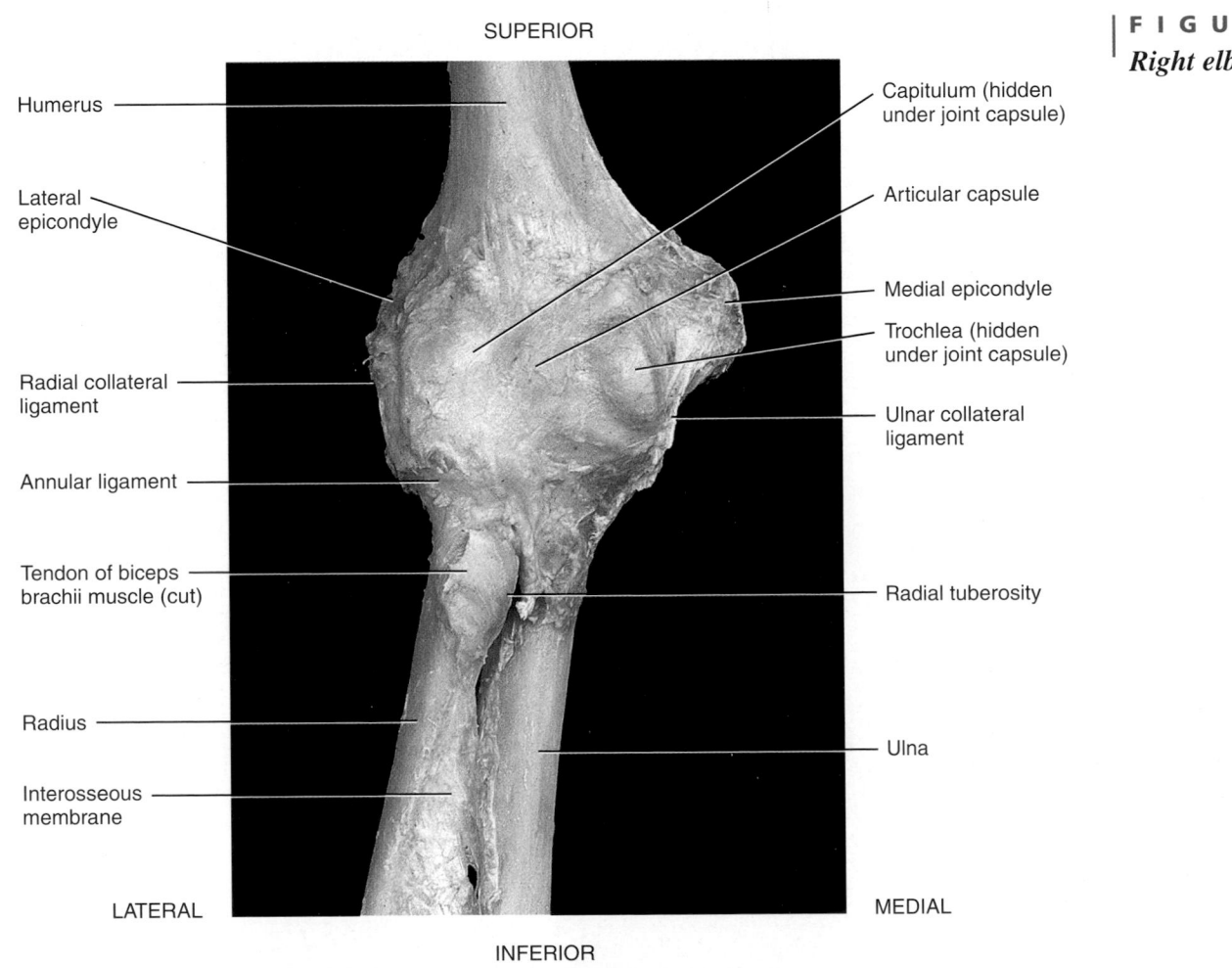

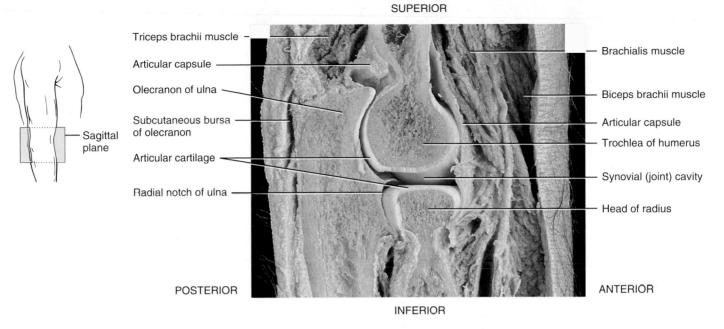

SUPERIOR

Triceps brachii muscle

Articular capsule

Olecranon of ulna

Subcutaneous bursa of olecranon

Articular cartilage

Radial notch of ulna

Brachialis muscle

Biceps brachii muscle

Articular capsule

Trochlea of humerus

Synovial (joint) cavity

Head of radius

POSTERIOR

ANTERIOR

INFERIOR

Sagittal plane

Sagittal section

FIGURE 4.4
Right elbow joint

SUPERIOR

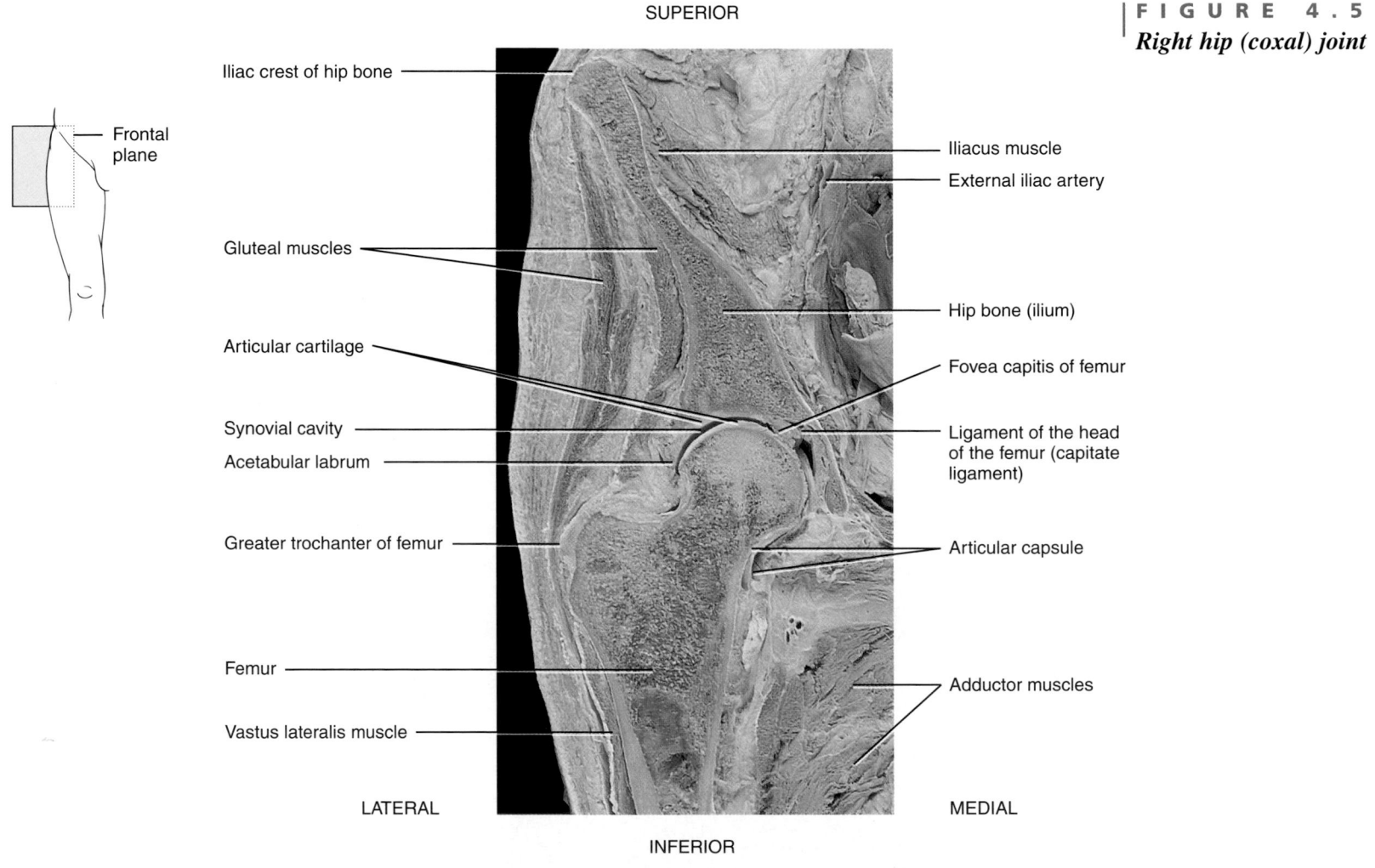

Iliac crest of hip bone

Iliacus muscle

External iliac artery

Frontal plane

Gluteal muscles

Hip bone (ilium)

Articular cartilage

Fovea capitis of femur

Synovial cavity

Ligament of the head of the femur (capitate ligament)

Acetabular labrum

Greater trochanter of femur

Articular capsule

Femur

Adductor muscles

Vastus lateralis muscle

LATERAL

MEDIAL

INFERIOR

Frontal section

FIGURE 4.5
Right hip (coxal) joint

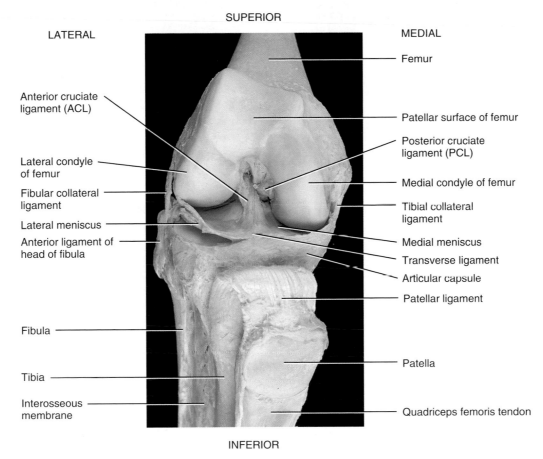

SUPERIOR

LATERAL

MEDIAL

Femur

Anterior cruciate
ligament (ACL)

Patellar surface of femur

Posterior cruciate
ligament (PCL)

Lateral condyle
of femur

Medial condyle of femur

Fibular collateral
ligament

Tibial collateral
ligament

Lateral meniscus

Medial meniscus

Anterior ligament of
head of fibula

Transverse ligament

Articular capsule

Patellar ligament

Fibula

Patella

Tibia

Interosseous
membrane

Quadriceps femoris tendon

INFERIOR

Anterior view

FIGURE 4.6
Right knee, flexed

SUPERIOR

Intercondylar
fossa

Femur

Anterior cruciate
ligament (ACL)

Lateral condyle of
femur (covered with
articular cartilage)

Medial condyle of femur
(covered with articular
cartilage)

Fibular (lateral) collateral
ligament

Tibial (medial) collateral
ligament

Lateral meniscus

Posterior cruciate
ligament (PCL)

Medial meniscus

Posterior ligament of
tibiofibular joint

Tibia

Fibula

Interosseus membrane

MEDIAL

LATERAL

INFERIOR

Posterior view

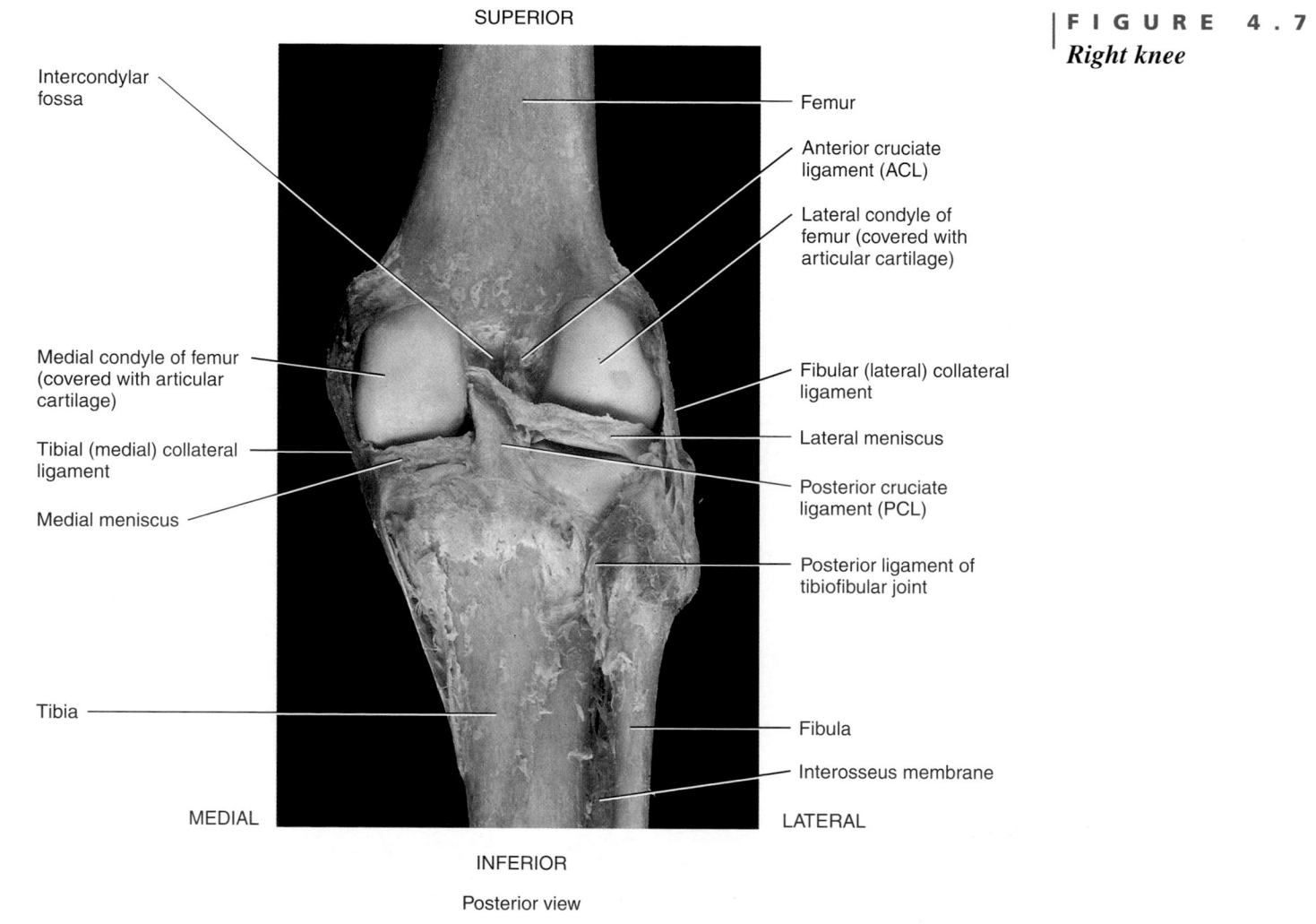

FIGURE 4.7
Right knee

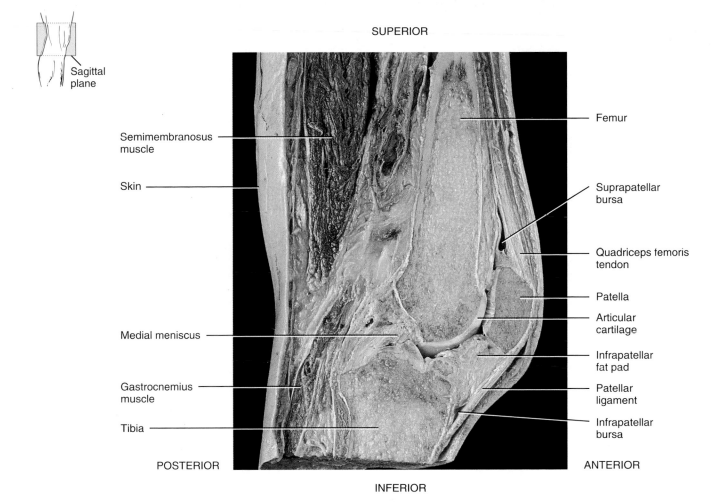

SUPERIOR

Semimembranosus
muscle

Skin

Medial meniscus

Gastrocnemius
muscle

Tibia

Femur

Suprapatellar
bursa

Quadriceps femoris
tendon

Patella

Articular
cartilage

Infrapatellar
fat pad

Patellar
ligament

Infrapatellar
bursa

POSTERIOR

ANTERIOR

INFERIOR

Sagittal section

Sagittal
plane

FIGURE 4.8
Right knee

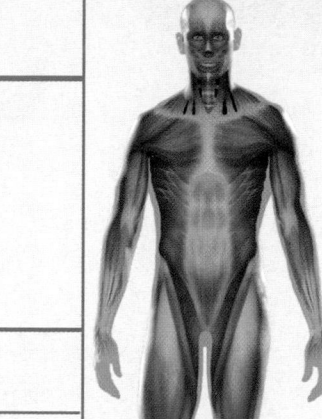

TABLE 5.1 | *Characteristics Used to Name Muscles*

DIRECTION: Orientation of muscle fascicles relative to the midline

Name	Meaning	Example
Rectus	parallel to midline	Rectus abdominis
Transverse	perpendicular to midline	Transversus abdominis
Oblique	diagonal to midline	External oblique

SIZE: Relative size of the muscle

Name	Meaning	Example
Maximus	largest	Gluteus maximus
Minimus	smallest	Gluteus minimus
Longus	longest	Adductor longus
Brevis	shortest	Fibularis brevis
Latissimus	widest	Latissimus dorsi
Longissimus	longest	Longissimus capitis
Magnus	large	Adductor magnus
Major	larger	Pectoralis major
Minor	smaller	Pectoralis minor
Vastus	great	Vastus lateralis

SHAPE: Relative shape of the muscle

Name	Meaning	Example
Deltoid	triangular	Deltoid
Trapezius	trapezoid	Trapezius
Serratus	saw-toothed	Serratus anterior
Rhomboid	diamond-shaped	Rhomboid major
Orbicularis	circular	Orbicularis oculi
Pectinate	comblike	Pectineus
Piriformis	pear-shaped	Piriformis
Platys	flat	Platysma
Quadratus	square	Quadratus femoris
Gracilis	slender	Gracilis

ACTION: Principal action of the muscle

Name	Meaning	Example
Flexor	decreases joint angle	Flexor carpi radialis
Extensor	increases joint angle	Extensor carpi ulnaris
Abductor	moves bone away from midline	Abductor pollicis longus
Adductor	moves bone closer to midline	Adductor longus
Levator	produces superior movement	Levator scapulae
Depressor	produces inferior movement	Depressor labii inferioris
Supinator	turns palm superiorly or anteriorly	Supinator
Pronator	turns palm inferiorly or posteriorly	Pronator teres
Sphincter	decreases size of opening	External anal sphincter
Tensor	makes a body part rigid	Tensor fasciae latae
Rotator	moves bone around longitudinal axis	Obturator externus

NUMBER OF ORIGINS: Number of tendons of origin

Name	Meaning	Example
Biceps	Two origins	Biceps brachii
Triceps	Three origins	Triceps brachii
Quadriceps	Four origins	Quadriceps femoris

LOCATION: Structure near which a muscle is found

Example

Frontalis, a muscle near the frontal bone; tibialis anterior, a muscle near the front of the tibia.

ORIGIN AND INSERTION: Sites where muscle originates and inserts

Example

Sternocleidomastoid, originates on the sternum and clavicle and inserts on mastoid process of the temporal bone; stylohyoid, originates on the styloid process of the temporal bone and inserts on the hyoid bone.

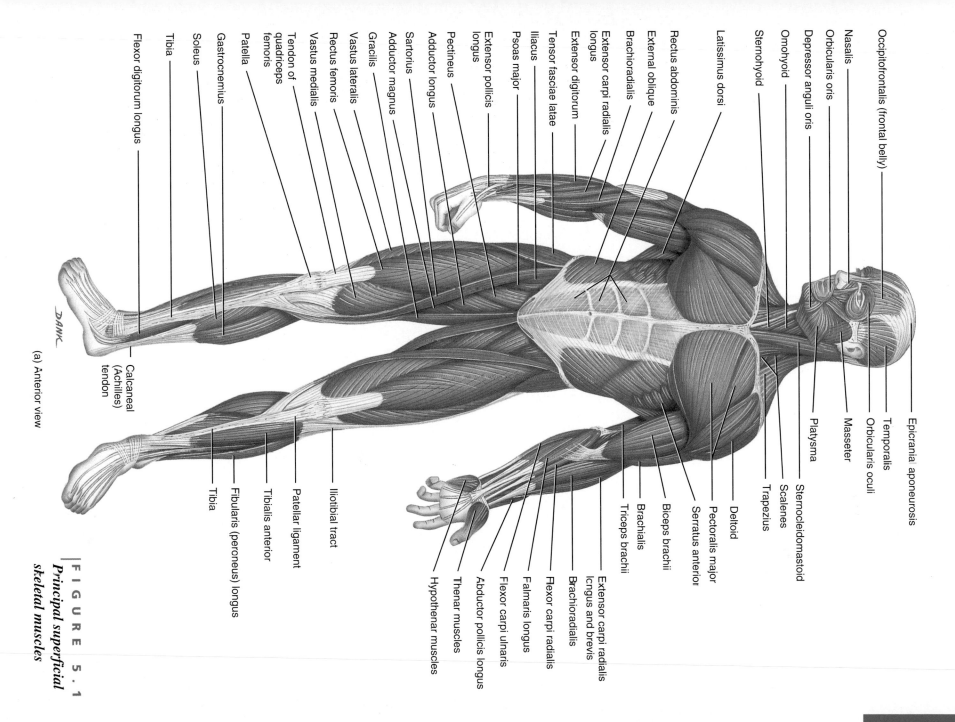

Flexor digitorum longus

Tibia

Soleus

Gastrocnemius

Patella

Tendon of quadriceps femoris

Vastus medialis

Rectus femoris

Vastus lateralis

Gracilis

Adductor magnus

Sartorius

Adductor longus

Pectineus

Extensor pollicis longus

Psoas major

Iliacus

Tensor fasciae latae

Extensor digitorum

Extensor carpi radialis longus

Brachioradialis

External oblique

Rectus abdominis

Latissimus dorsi

Sternohyoid

Omohyoid

Depressor anguli oris

Orbicularis oris

Nasalis

Occipitofrontalis (frontal belly)

Epicranial aponeurosis

Temporalis

Orbicularis oculi

Masseter

Platysma

Sternocleidomastoid

Scalenes

Trapezius

Deltoid

Pectoralis major

Serratus anterior

Biceps brachii

Brachialis

Triceps brachii

Extensor carpi radialis longus and brevis

Brachioradialis

Flexor carpi radialis

Falmaris longus

Flexor carpi ulnaris

Abductor pollicis longus

Thenar muscles

Hypothenar muscles

Iliotibial tract

Patellar ligament

Tibialis anterior

Fibularis (peroneus) longus

Tibia

Calcaneal (Achilles) tendon

(a) Anterior view

DANK

FIGURE 5.1
Principal superficial skeletal muscles

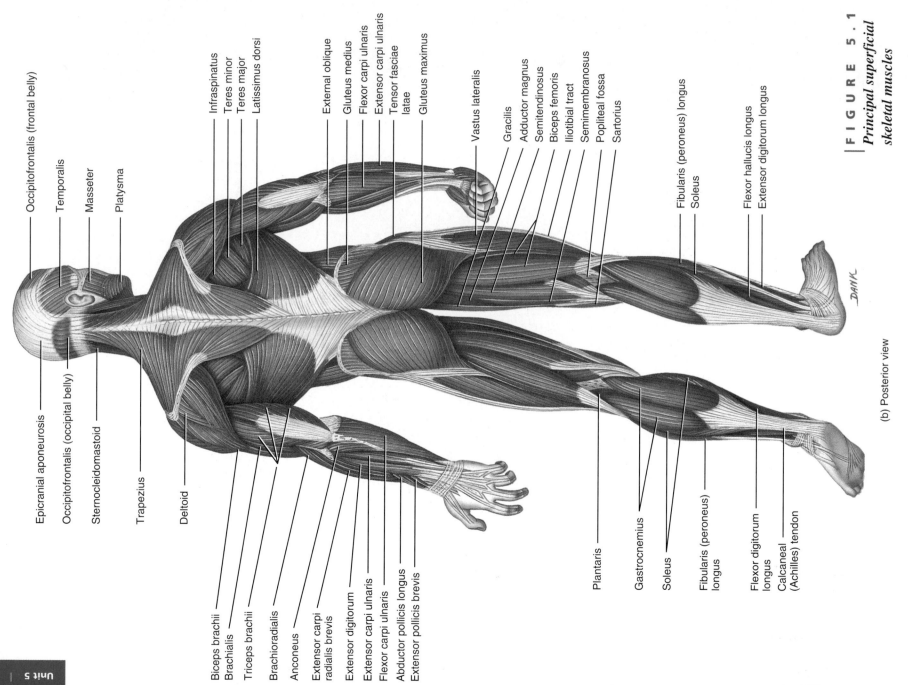

Occipitofrontalis (frontal belly)

Temporalis

Masseter

Platysma

Epicranial aponeurosis

Occipitofrontalis (occipital belly)

Sternocleidomastoid

Trapezius

Deltoid

Biceps brachii
Brachialis
Triceps brachii
Brachioradialis
Anconeus
Extensor carpi radialis brevis
Extensor digitorum
Extensor carpi ulnaris
Flexor carpi ulnaris
Abductor pollicis longus
Extensor pollicis brevis

Infraspinatus
Teres minor
Teres major
Latissimus dorsi

External oblique
Gluteus medius
Flexor carpi ulnaris
Extensor carpi ulnaris
Tensor fasciae latae
Gluteus maximus

Vastus lateralis

Gracilis
Adductor magnus
Semitendinosus
Biceps femoris
Iliotibial tract
Semimembranosus
Popliteal fossa
Sartorius

Fibularis (peroneus) longus
Soleus

Flexor hallucis longus
Extensor digitorum longus

Plantaris

Gastrocnemius

Soleus

Fibularis (peroneus) longus

Flexor digitorum longus

Calcaneal (Achilles) tendon

DAHL

F I G U R E   5 . 1
Principal superficial skeletal muscles

(b) Posterior view

FIGURE 5.2
Muscles of facial expression

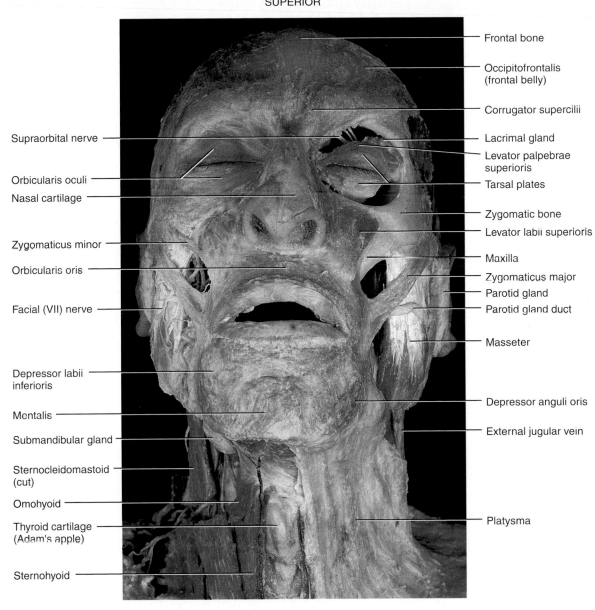

Frontal bone

Occipitofrontalis (frontal belly)

Corrugator supercilii

Supraorbital nerve

Lacrimal gland

Levator palpebrae superioris

Orbicularis oculi

Nasal cartilage

Tarsal plates

Zygomatic bone

Levator labii superioris

Zygomaticus minor

Maxilla

Orbicularis oris

Zygomaticus major

Parotid gland

Facial (VII) nerve

Parotid gland duct

Masseter

Depressor labii inferioris

Mentalis

Depressor anguli oris

Submandibular gland

External jugular vein

Sternocleidomastoid (cut)

Omohyoid

Thyroid cartilage (Adam's apple)

Platysma

Sternohyoid

Anterior view

FIGURE 5.3
Muscles of facial expression

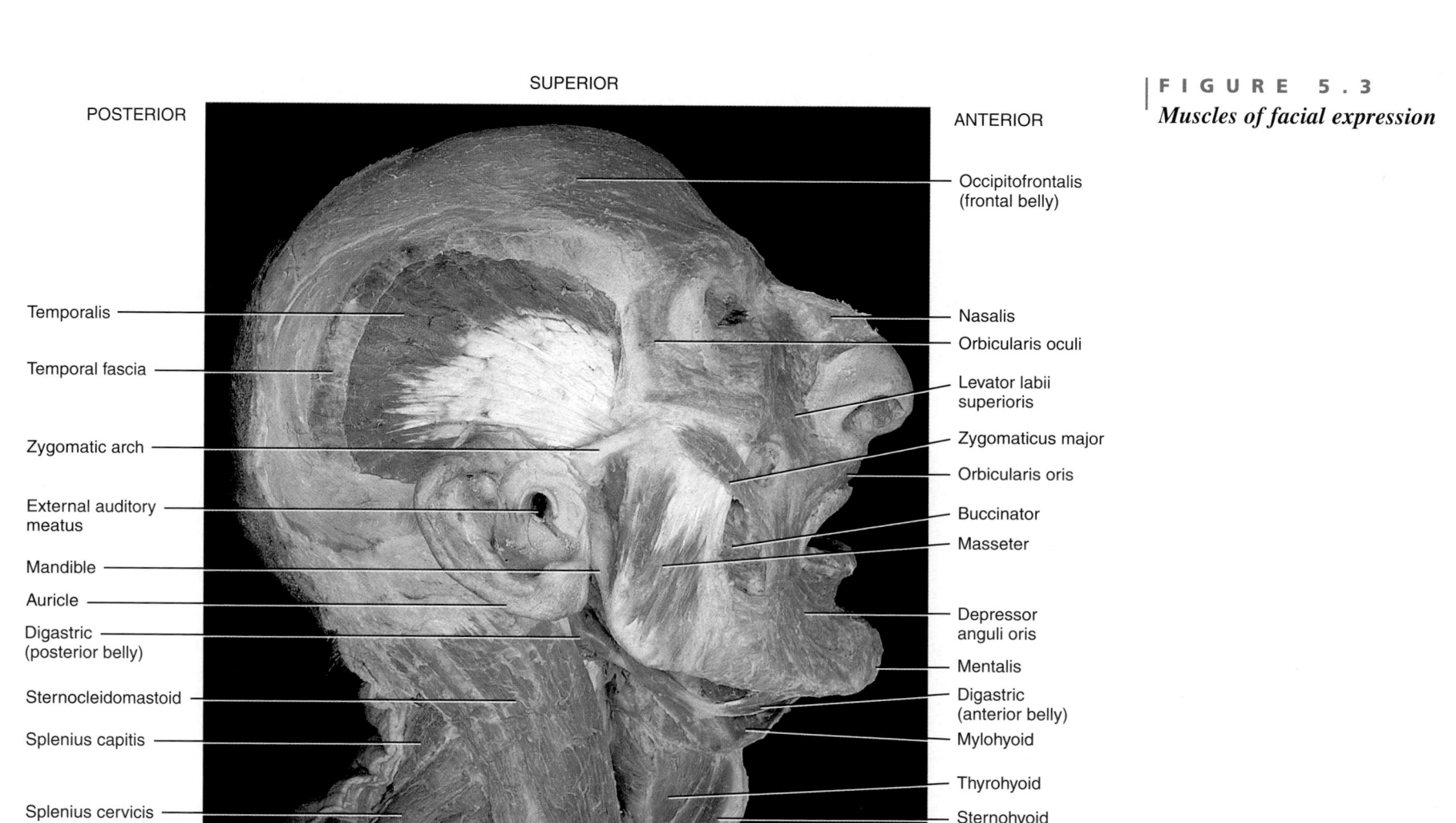

SUPERIOR

POSTERIOR

ANTERIOR

Occipitofrontalis
(frontal belly)

Temporalis

Nasalis

Orbicularis oculi

Temporal fascia

Levator labii
superioris

Zygomatic arch

Zygomaticus major

Orbicularis oris

External auditory
meatus

Buccinator

Masseter

Mandible

Auricle

Depressor
anguli oris

Digastric
(posterior belly)

Mentalis

Sternocleidomastoid

Digastric
(anterior belly)

Splenius capitis

Mylohyoid

Thyrohyoid

Splenius cervicis

Sternohyoid

Omohyoid

INFERIOR

Right lateral view

SUPERIOR

POSTERIOR

ANTERIOR

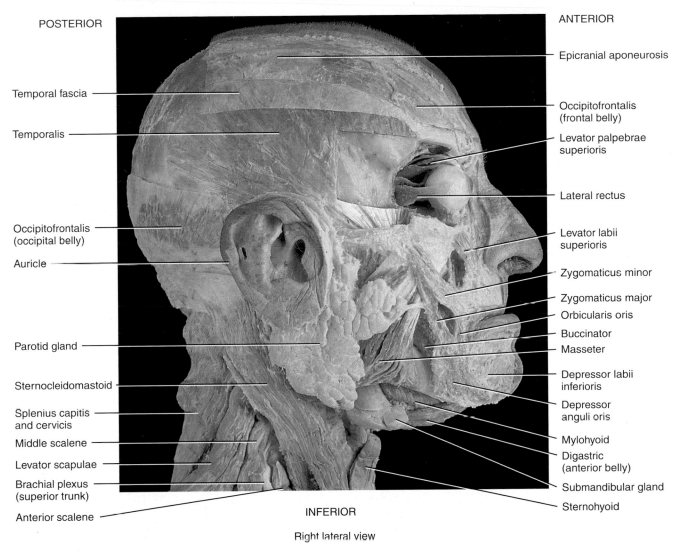

Temporal fascia

Temporalis

Occipitofrontalis
(occipital belly)

Auricle

Parotid gland

Sternocleidomastoid

Splenius capitis
and cervicis

Middle scalene

Levator scapulae

Brachial plexus
(superior trunk)

Anterior scalene

Epicranial aponeurosis

Occipitofrontalis
(frontal belly)

Levator palpebrae
superioris

Lateral rectus

Levator labii
superioris

Zygomaticus minor

Zygomaticus major

Orbicularis oris

Buccinator

Masseter

Depressor labii
inferioris

Depressor
anguli oris

Mylohyoid

Digastric
(anterior belly)

Submandibular gland

Sternohyoid

INFERIOR

Right lateral view

FIGURE 5.4
Muscles of facial expression

SUPERIOR

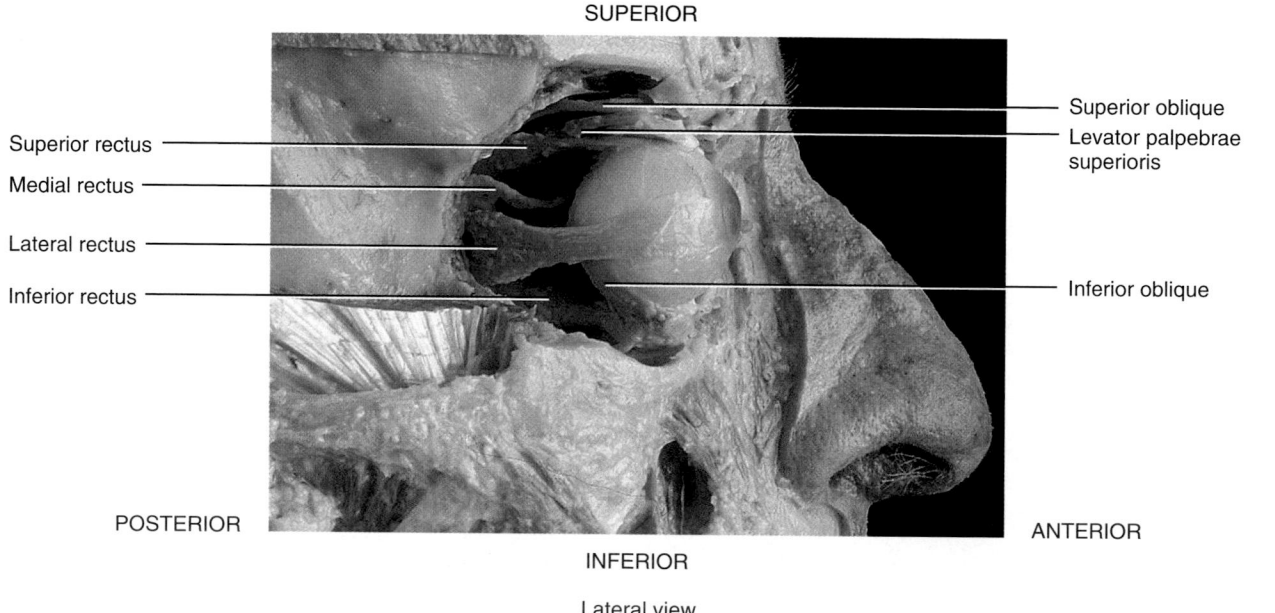

Superior rectus

Medial rectus

Lateral rectus

Inferior rectus

Superior oblique

Levator palpebrae
superioris

Inferior oblique

POSTERIOR

ANTERIOR

INFERIOR

Lateral view

FIGURE 5.5
Extrinsic muscles of the right eyeball

SUPERIOR

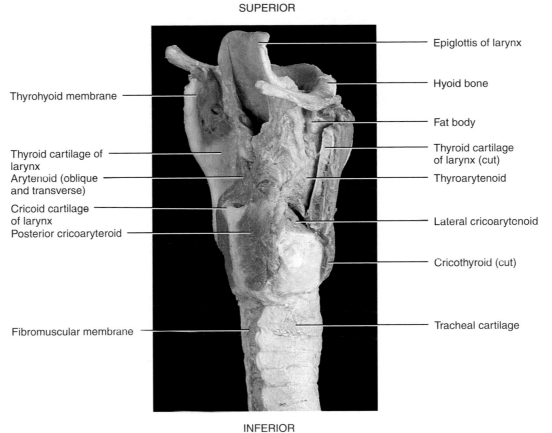

Thyrohyoid membrane

Thyroid cartilage of
larynx
Arytenoid (oblique
and transverse)
Cricoid cartilage
of larynx
Posterior cricoaryteroid

Fibromuscular membrane

Epiglottis of larynx

Hyoid bone

Fat body

Thyroid cartilage
of larynx (cut)

Thyroarytenoid

Lateral cricoarytenoid

Cricothyroid (cut)

Tracheal cartilage

INFERIOR

Right posterolateral view

FIGURE 5.6
Muscles of the larynx

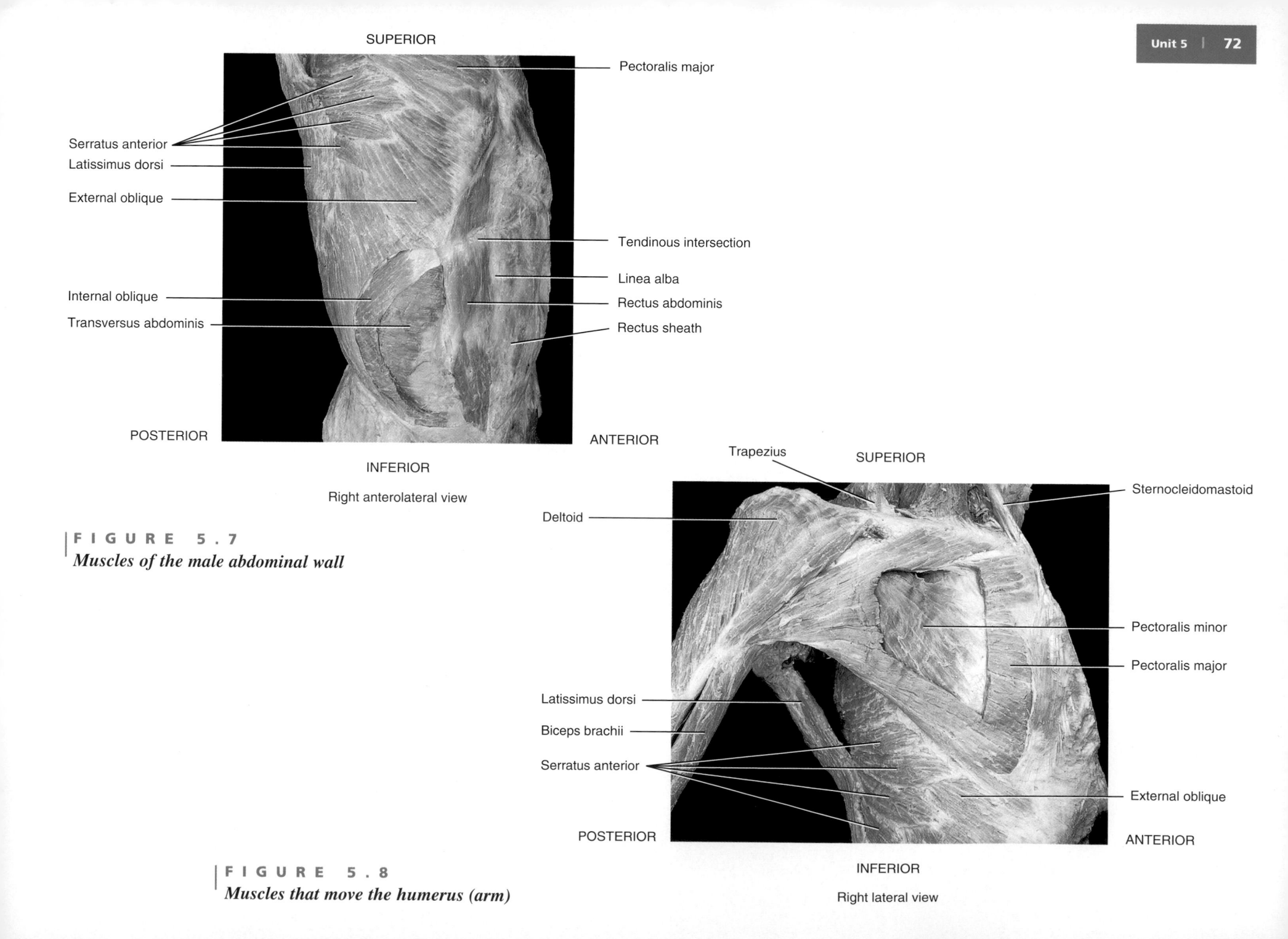

SUPERIOR

Pectoralis major

Serratus anterior

Latissimus dorsi

External oblique

Tendinous intersection

Linea alba

Internal oblique

Rectus abdominis

Transversus abdominis

Rectus sheath

POSTERIOR

ANTERIOR

INFERIOR

Right anterolateral view

FIGURE 5.7
Muscles of the male abdominal wall

Trapezius

SUPERIOR

Sternocleidomastoid

Deltoid

Pectoralis minor

Pectoralis major

Latissimus dorsi

Biceps brachii

Serratus anterior

External oblique

POSTERIOR

ANTERIOR

INFERIOR

FIGURE 5.8
Muscles that move the humerus (arm)

Right lateral view

SUPERIOR

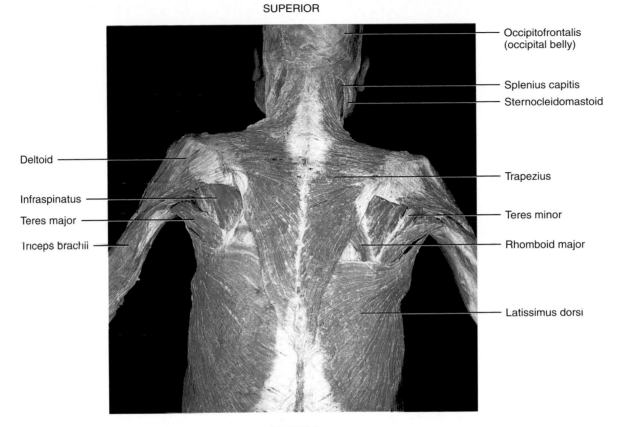

Occipitofrontalis
(occipital belly)

Splenius capitis

Sternocleidomastoid

Deltoid

Trapezius

Infraspinatus

Teres major

Teres minor

Triceps brachii

Rhomboid major

Latissimus dorsi

INFERIOR

Posterior view

FIGURE 5.9
Muscles that move the humerus (arm)

SUPERIOR

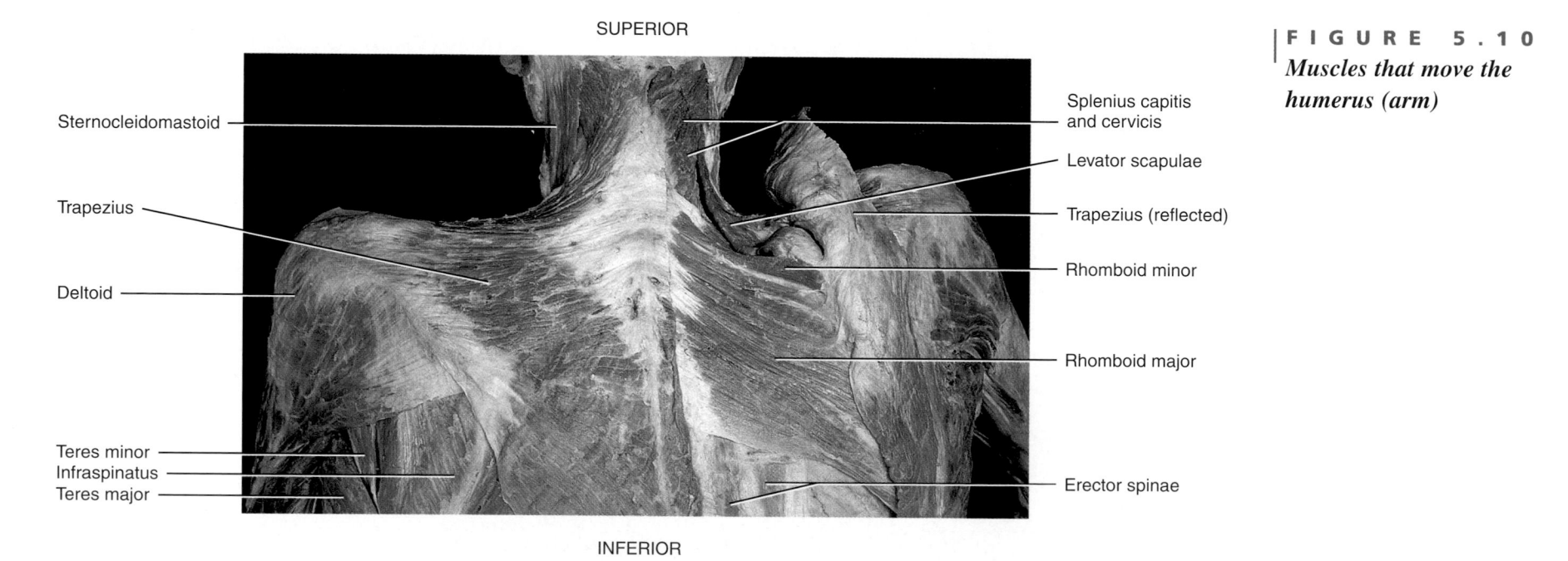

Sternocleidomastoid

Trapezius

Deltoid

Teres minor
Infraspinatus
Teres major

Splenius capitis
and cervicis

Levator scapulae

Trapezius (reflected)

Rhomboid minor

Rhomboid major

Erector spinae

INFERIOR

Posterior view

FIGURE 5.10
Muscles that move the humerus (arm)

SUPERIOR

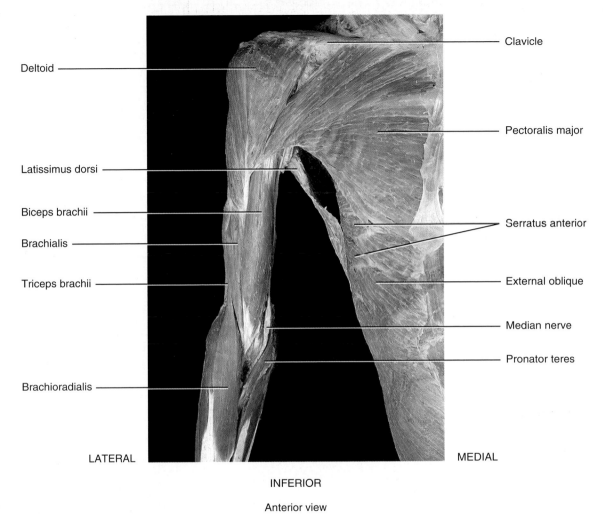

Deltoid

Latissimus dorsi

Biceps brachii

Brachialis

Triceps brachii

Brachioradialis

Clavicle

Pectoralis major

Serratus anterior

External oblique

Median nerve

Pronator teres

LATERAL

MEDIAL

INFERIOR

Anterior view

FIGURE 5.11

Muscles that move the right radius and ulna (forearm)

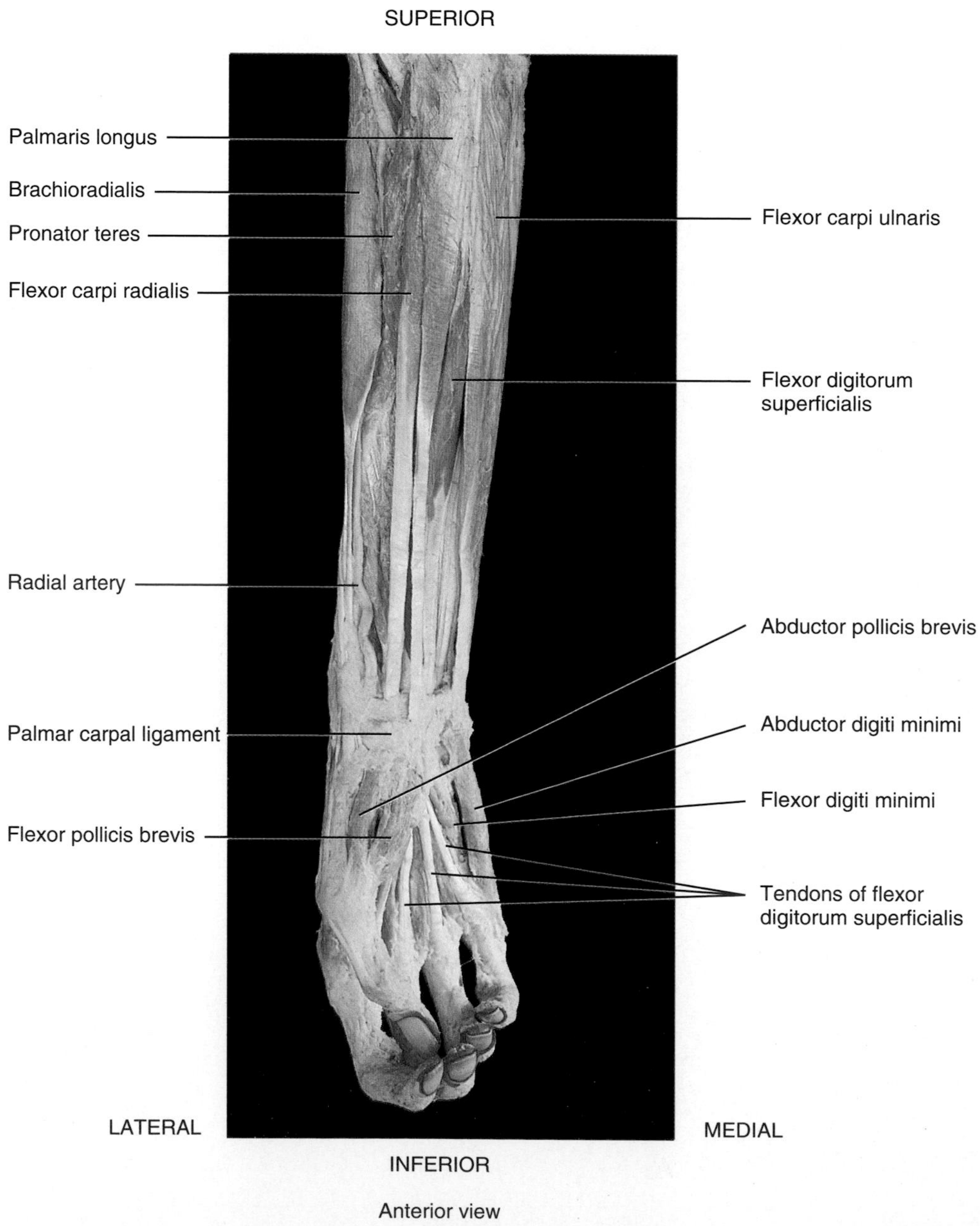

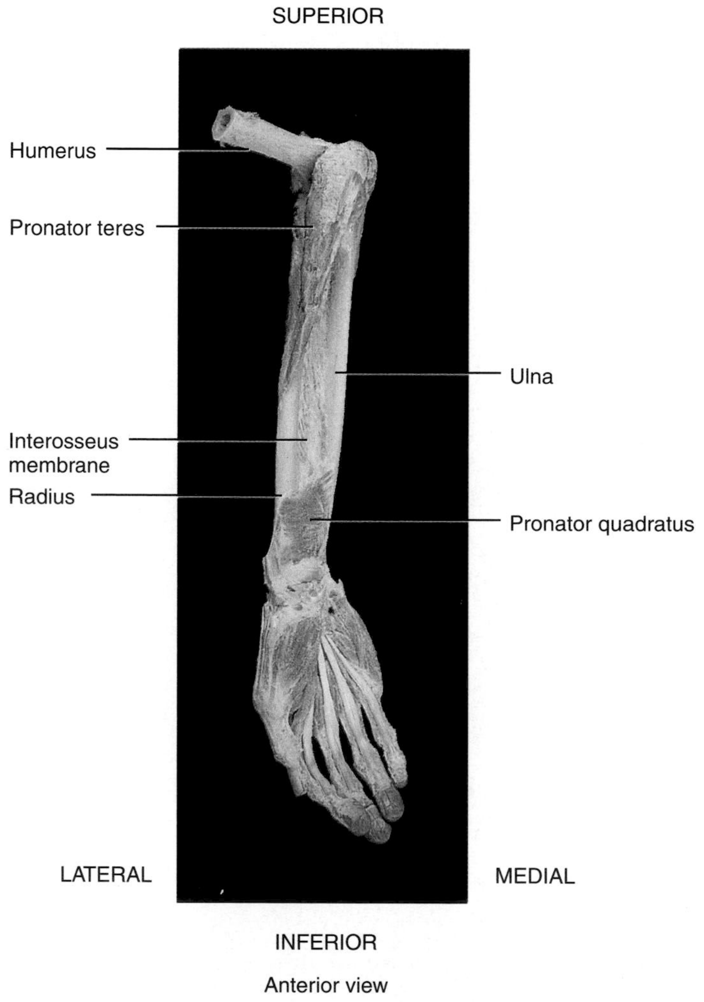

SUPERIOR

Palmaris longus

Brachioradialis

Pronator teres

Flexor carpi radialis

Flexor carpi ulnaris

Flexor digitorum superficialis

Radial artery

Abductor pollicis brevis

Abductor digiti minimi

Palmar carpal ligament

Flexor digiti minimi

Flexor pollicis brevis

Tendons of flexor digitorum superficialis

LATERAL   MEDIAL

INFERIOR

Anterior view

SUPERIOR

Humerus

Pronator teres

Interosseus membrane

Radius

Ulna

Pronator quadratus

LATERAL   MEDIAL

INFERIOR

Anterior view

FIGURE 5.12

Muscles that move the right wrist, hand, and fingers

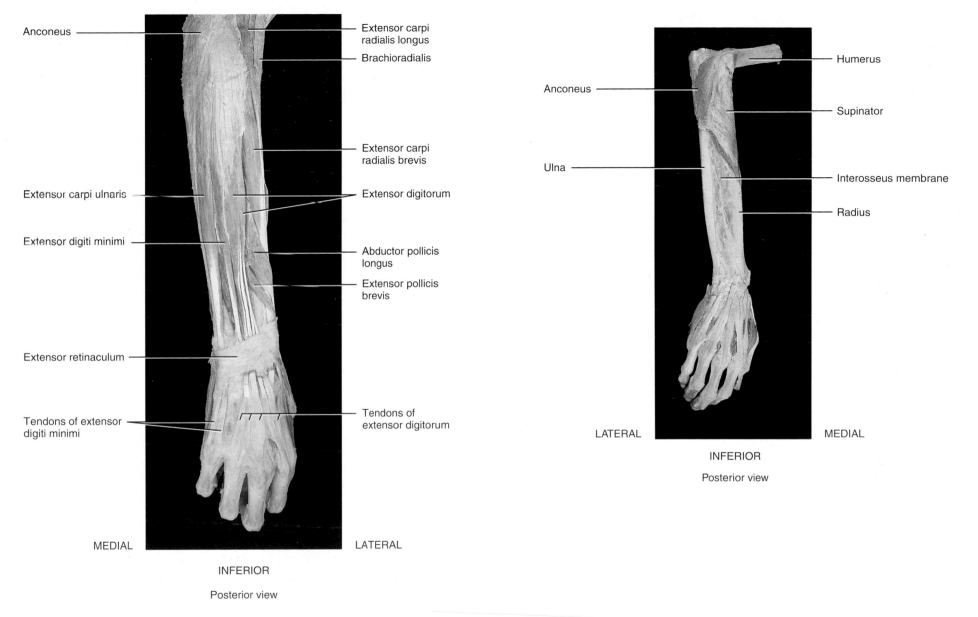

SUPERIOR

Anconeus

Extensor carpi radialis longus

Brachioradialis

Extensor carpi radialis brevis

Extensor carpi ulnaris

Extensor digitorum

Extensor digiti minimi

Abductor pollicis longus

Extensor pollicis brevis

Extensor retinaculum

Tendons of extensor digiti minimi

Tendons of extensor digitorum

MEDIAL

LATERAL

INFERIOR

Posterior view

SUPERIOR

Anconeus

Humerus

Supinator

Ulna

Interosseus membrane

Radius

LATERAL

MEDIAL

INFERIOR

Posterior view

FIGURE 5.13

Muscles that move the right wrist, hand, and fingers

SUPERIOR

FIGURE 5.14
Intrinsic muscles of the right hand

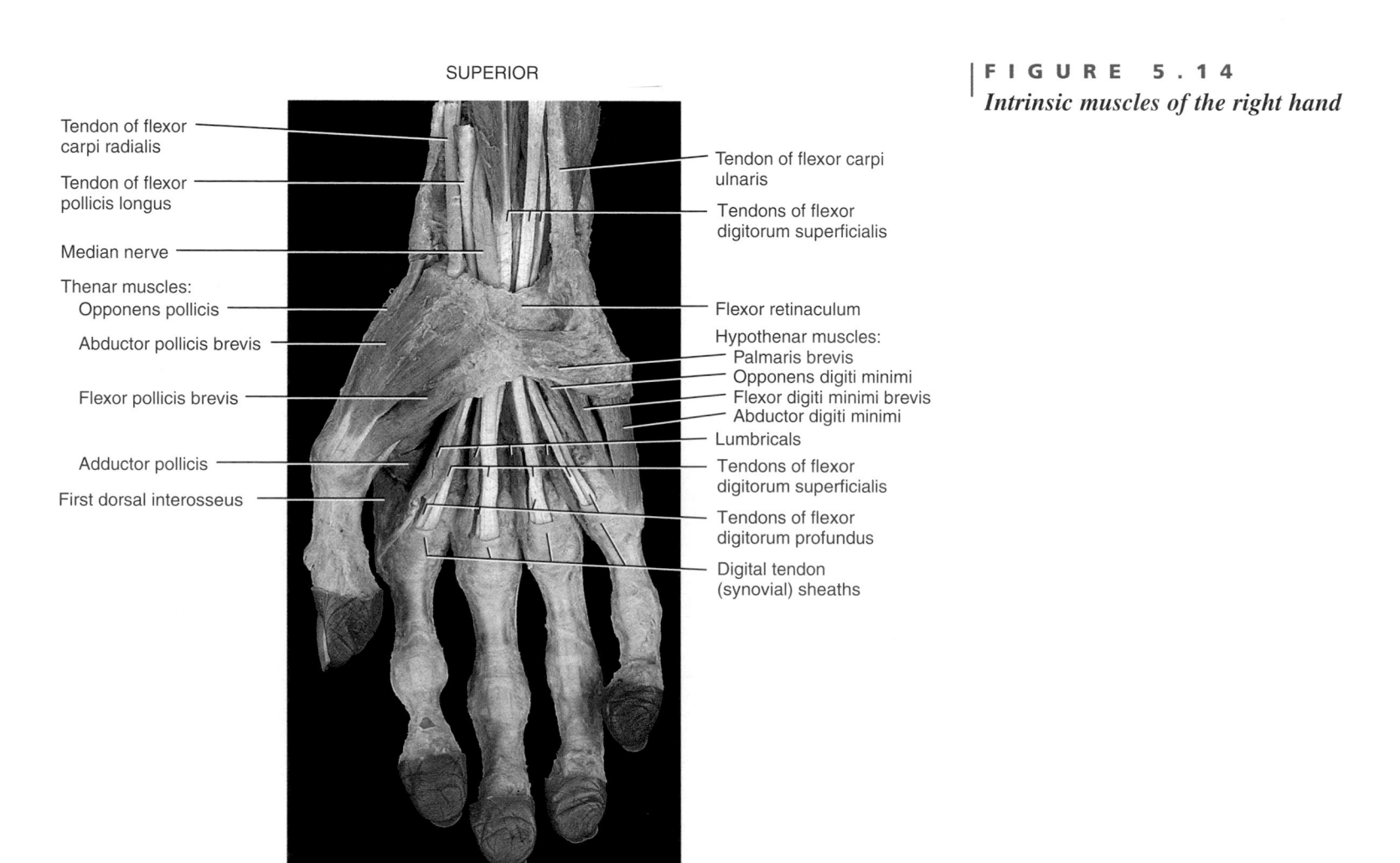

Tendon of flexor
carpi radialis

Tendon of flexor
pollicis longus

Median nerve

Thenar muscles:
  Opponens pollicis

  Abductor pollicis brevis

Flexor pollicis brevis

Adductor pollicis

First dorsal interosseus

Tendon of flexor carpi
ulnaris

Tendons of flexor
digitorum superficialis

Flexor retinaculum

Hypothenar muscles:
  Palmaris brevis
  Opponens digiti minimi
  Flexor digiti minimi brevis
  Abductor digiti minimi
Lumbricals

Tendons of flexor
digitorum superficialis

Tendons of flexor
digitorum profundus

Digital tendon
(synovial) sheaths

LATERAL

MEDIAL

INFERIOR

Anterior view

SUPERIOR

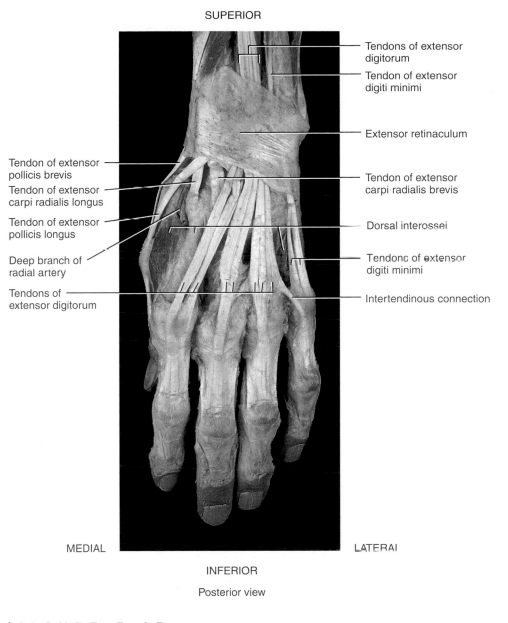

Tendons of extensor
digitorum

Tendon of extensor
digiti minimi

Extensor retinaculum

Tendon of extensor
pollicis brevis

Tendon of extensor
carpi radialis longus

Tendon of extensor
pollicis longus

Deep branch of
radial artery

Tendons of
extensor digitorum

Tendon of extensor
carpi radialis brevis

Dorsal interossei

Tendons of extensor
digiti minimi

Intertendinous connection

MEDIAL

LATERAL

INFERIOR

Posterior view

FIGURE 5.15
Intrinsic muscles of the left hand

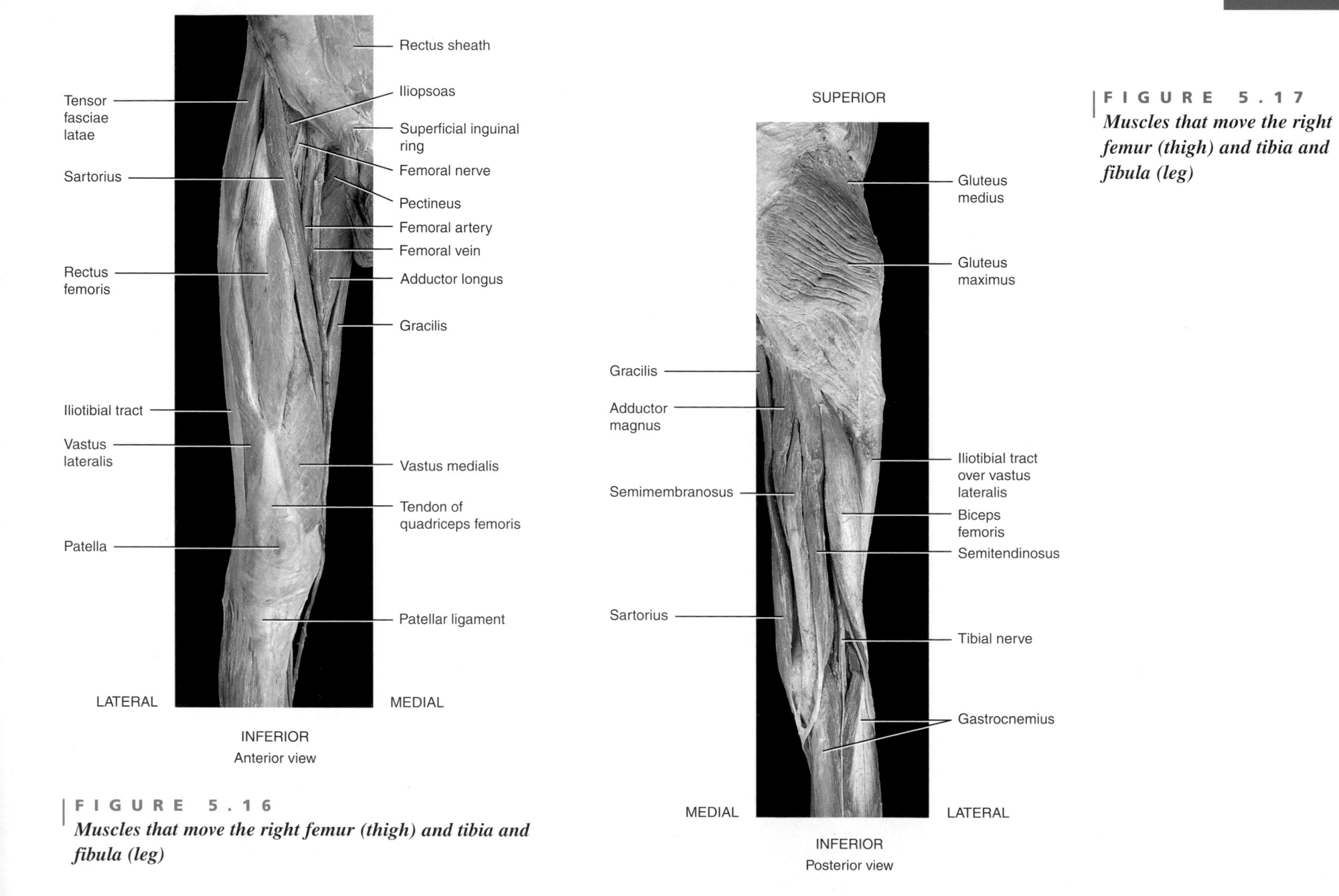

SUPERIOR

Rectus sheath

Tensor fasciae latae

Iliopsoas

Superficial inguinal ring

Femoral nerve

Sartorius

Pectineus

Femoral artery

Femoral vein

Rectus femoris

Adductor longus

Gracilis

Iliotibial tract

Vastus lateralis

Vastus medialis

Tendon of quadriceps femoris

Patella

Patellar ligament

LATERAL

MEDIAL

INFERIOR

Anterior view

FIGURE 5.16

Muscles that move the right femur (thigh) and tibia and fibula (leg)

FIGURE 5.17

Muscles that move the right femur (thigh) and tibia and fibula (leg)

SUPERIOR

Gluteus medius

Gluteus maximus

Gracilis

Adductor magnus

Iliotibial tract over vastus lateralis

Semimembranosus

Biceps femoris

Semitendinosus

Sartorius

Tibial nerve

Gastrocnemius

MEDIAL

LATERAL

INFERIOR

Posterior view

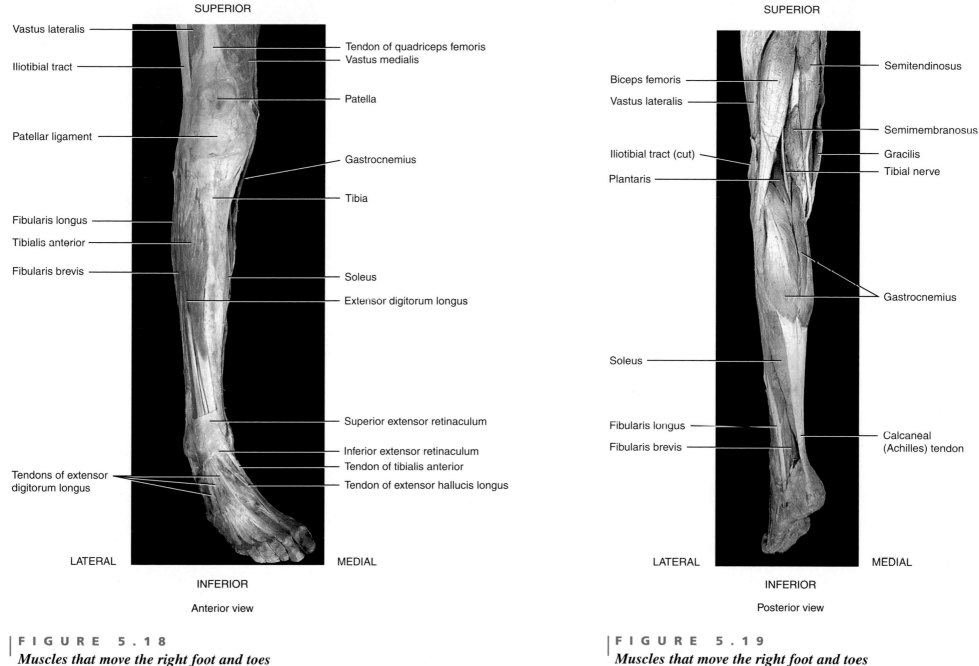

SUPERIOR

Vastus lateralis

Iliotibial tract

Patellar ligament

Fibularis longus

Tibialis anterior

Fibularis brevis

Tendons of extensor
digitorum longus

Tendon of quadriceps femoris
Vastus medialis

Patella

Gastrocnemius

Tibia

Soleus

Extensor digitorum longus

Superior extensor retinaculum

Inferior extensor retinaculum
Tendon of tibialis anterior
Tendon of extensor hallucis longus

LATERAL

MEDIAL

INFERIOR

Anterior view

FIGURE 5.18

Muscles that move the right foot and toes

SUPERIOR

Biceps femoris

Vastus lateralis

Iliotibial tract (cut)

Plantaris

Soleus

Fibularis longus

Fibularis brevis

Semitendinosus

Semimembranosus

Gracilis

Tibial nerve

Gastrocnemius

Calcaneal
(Achilles) tendon

LATERAL

MEDIAL

INFERIOR

Posterior view

FIGURE 5.19

Muscles that move the right foot and toes

UNIT SIX | *The Cardiovascular System*

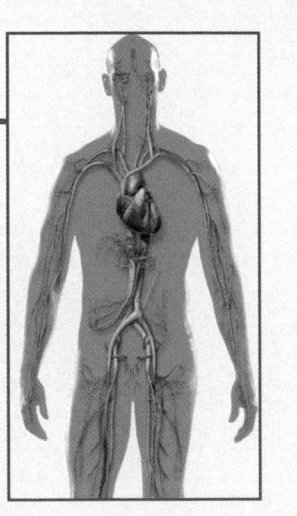

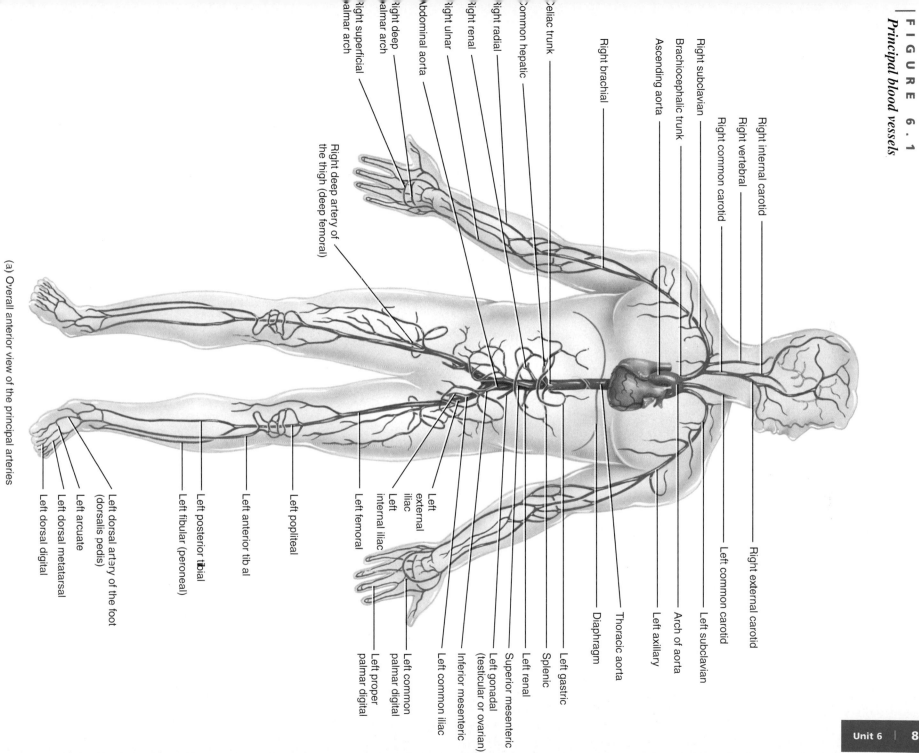

Right internal carotid

Right vertebral

Right common carotid

Right subclavian

Brachiocephalic trunk

Ascending aorta

Right brachial

Celiac trunk

Common hepatic

Right radial

Right renal

Right ulnar

Abdominal aorta

Right deep artery of
the thigh (deep femoral)

Right deep
palmar arch

Right superficial
palmar arch

Right dorsal digital

Left dorsal metatarsal

Left arcuate

Left dorsal artery of the foot
(dorsalis pedis)

Left fibular (peroneal)

Left posterior tibial

Left anterior tibial

Left popliteal

Left femoral

Left
internal iliac

Left
external
iliac

Left common iliac

Left proper
palmar digital

Left common
palmar digital

Inferior mesenteric

Left gonadal
(testicular or ovarian)

Superior mesenteric

Left renal

Splenic

Left gastric

Diaphragm

Thoracic aorta

Left axillary

Arch of aorta

Left subclavian

Left common carotid

Right external carotid

(a) Overall anterior view of the principal arteries

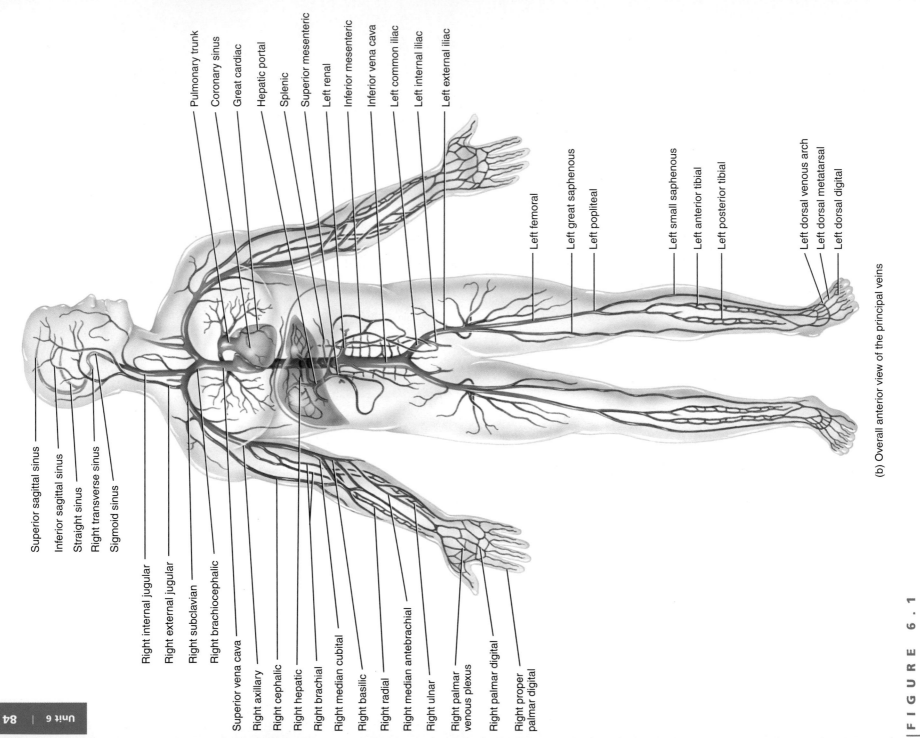

Superior sagittal sinus
Inferior sagittal sinus
Straight sinus
Right transverse sinus
Sigmoid sinus

Right internal jugular
Right external jugular
Right subclavian
Right brachiocephalic

Superior vena cava
Right axillary
Right cephalic
Right hepatic
Right brachial
Right median cubital
Right basilic
Right radial
Right median antebrachial
Right ulnar
Right palmar venous plexus
Right palmar digital
Right proper palmar digital

Pulmonary trunk
Coronary sinus
Great cardiac
Hepatic portal
Splenic
Superior mesenteric
Left renal
Inferior mesenteric
Inferior vena cava
Left common iliac
Left internal iliac
Left external iliac

Left femoral
Left great saphenous
Left popliteal

Left small saphenous
Left anterior tibial
Left posterior tibial

Left dorsal venous arch
Left dorsal metatarsal
Left dorsal digital

(b) Overall anterior view of the principal veins

FIGURE 6.1 *Principal blood vessels, continued*

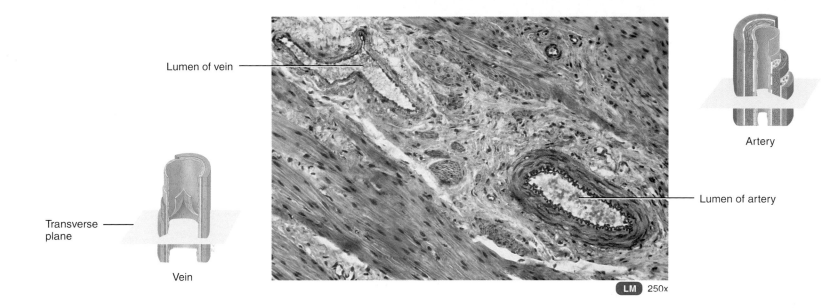

Lumen of vein

Transverse plane

Vein

Artery

Lumen of artery

LM 250x

(a) Transverse section through a vein and its accompanying artery

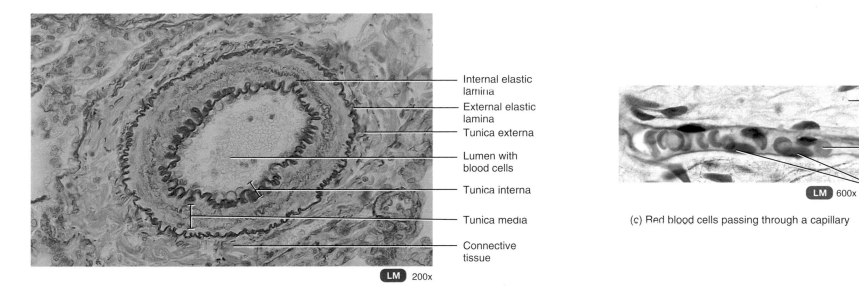

Internal elastic lamina

External elastic lamina

Tunica externa

Lumen with blood cells

Tunica interna

Tunica media

Connective tissue

LM 200x

(b) Transverse section through an artery

Connective tissue

Red blood cell

Capillary endothelial cells

LM 600x

(c) Red blood cells passing through a capillary

F I G U R E   6 . 2   *Histology of blood vessels*

FIGURE 6.3
Heart

SUPERIOR

Brachiocephalic vein

Brachiocephalic trunk

Superior vena cava

Ascending aorta

Right auricle of
right atrium

Adipose tissue over
right ventricle

Right ventricle

Left subclavian artery

Left common carotid artery

Arch of aorta

Ligamentum arteriosum

Left pulmonary artery

Pulmonary trunk

Left auricle of left atrium

Anterior interventricular
sulcus and vessels

Left ventricle

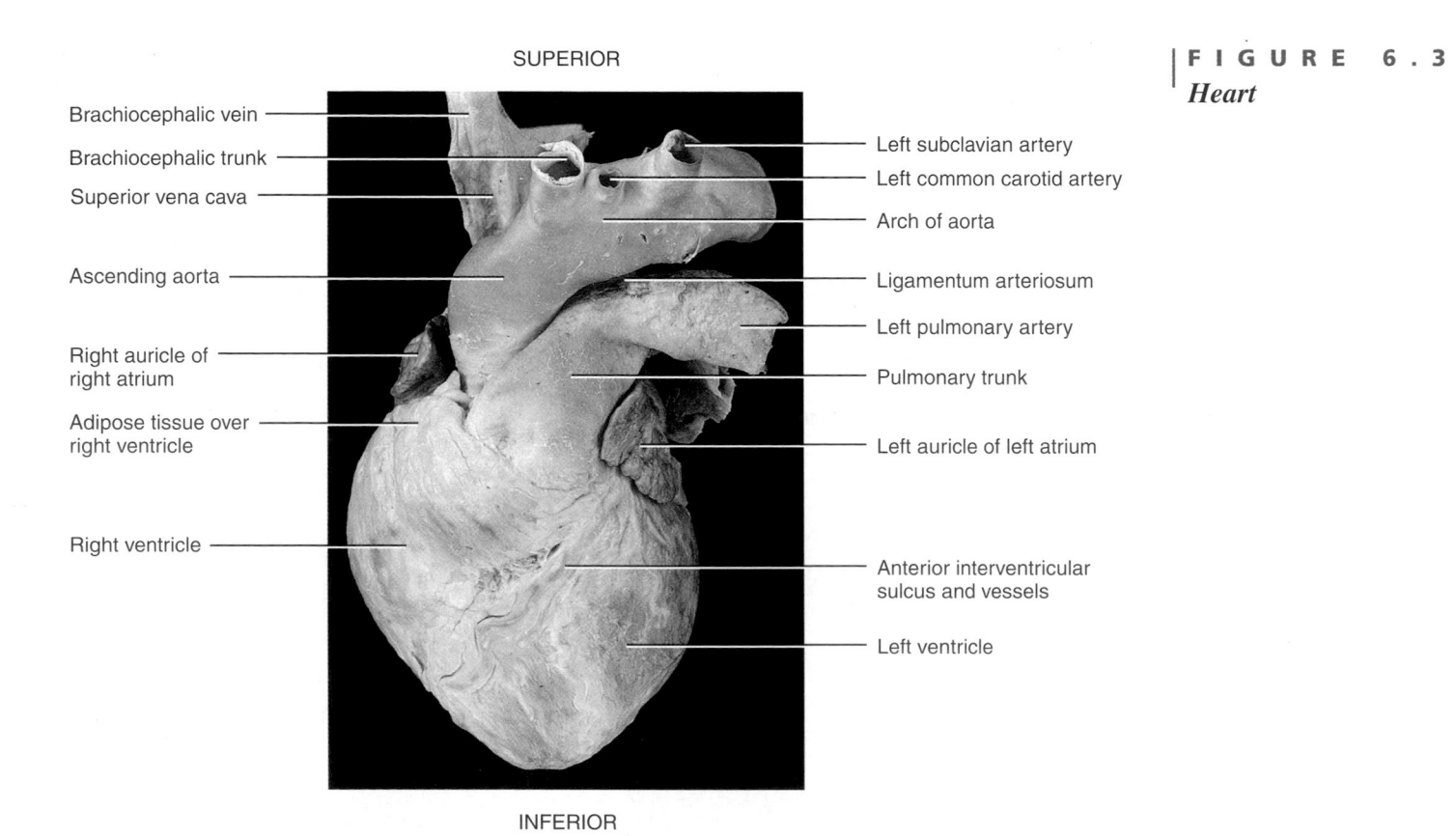

INFERIOR

(a) Anterior view of human heart

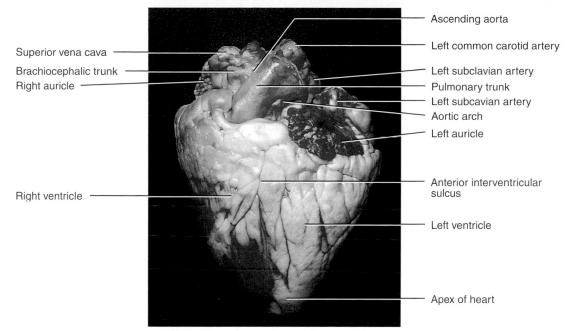

Superior vena cava

Brachiocephalic trunk

Right auricle

Ascending aorta

Left common carotid artery

Left subclavian artery

Pulmonary trunk

Left subcavian artery

Aortic arch

Left auricle

Right ventricle

Anterior interventricular sulcus

Left ventricle

Apex of heart

(b) Anterior view of sheep heart

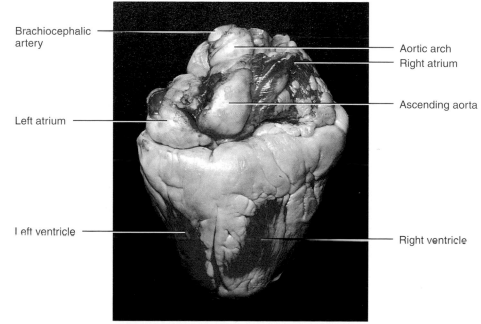

Brachiocephalic artery

Left atrium

Left ventricle

Aortic arch

Right atrium

Ascending aorta

Right ventricle

(c) Posterior view of sheep heart

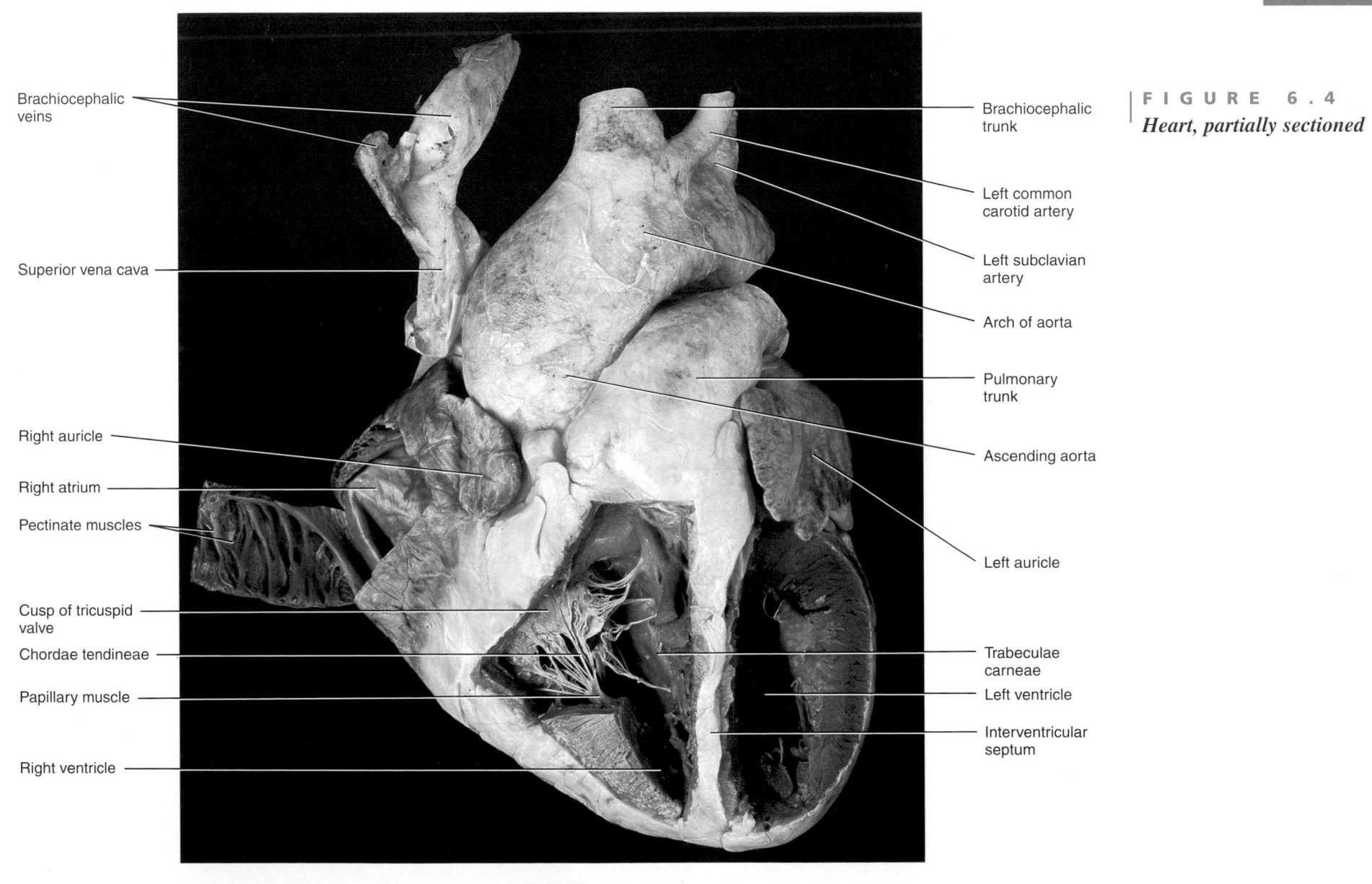

Brachiocephalic veins

Superior vena cava

Right auricle

Right atrium

Pectinate muscles

Cusp of tricuspid valve

Chordae tendineae

Papillary muscle

Right ventricle

Brachiocephalic trunk

Left common carotid artery

Left subclavian artery

Arch of aorta

Pulmonary trunk

Ascending aorta

Left auricle

Trabeculae carneae

Left ventricle

Interventricular septum

INFERIOR

(a) Anterior view of human heart

FIGURE 6.4
Heart, partially sectioned

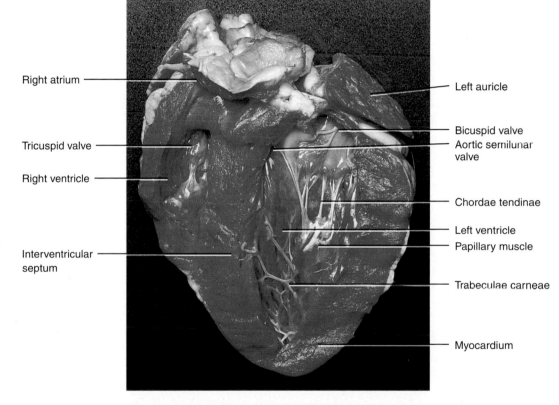

Right atrium

Tricuspid valve

Right ventricle

Interventricular
septum

Left auricle

Bicuspid valve

Aortic semilunar
valve

Chordae tendinae

Left ventricle

Papillary muscle

Trabeculae carneae

Myocardium

(b) Anterior view of sheep heart

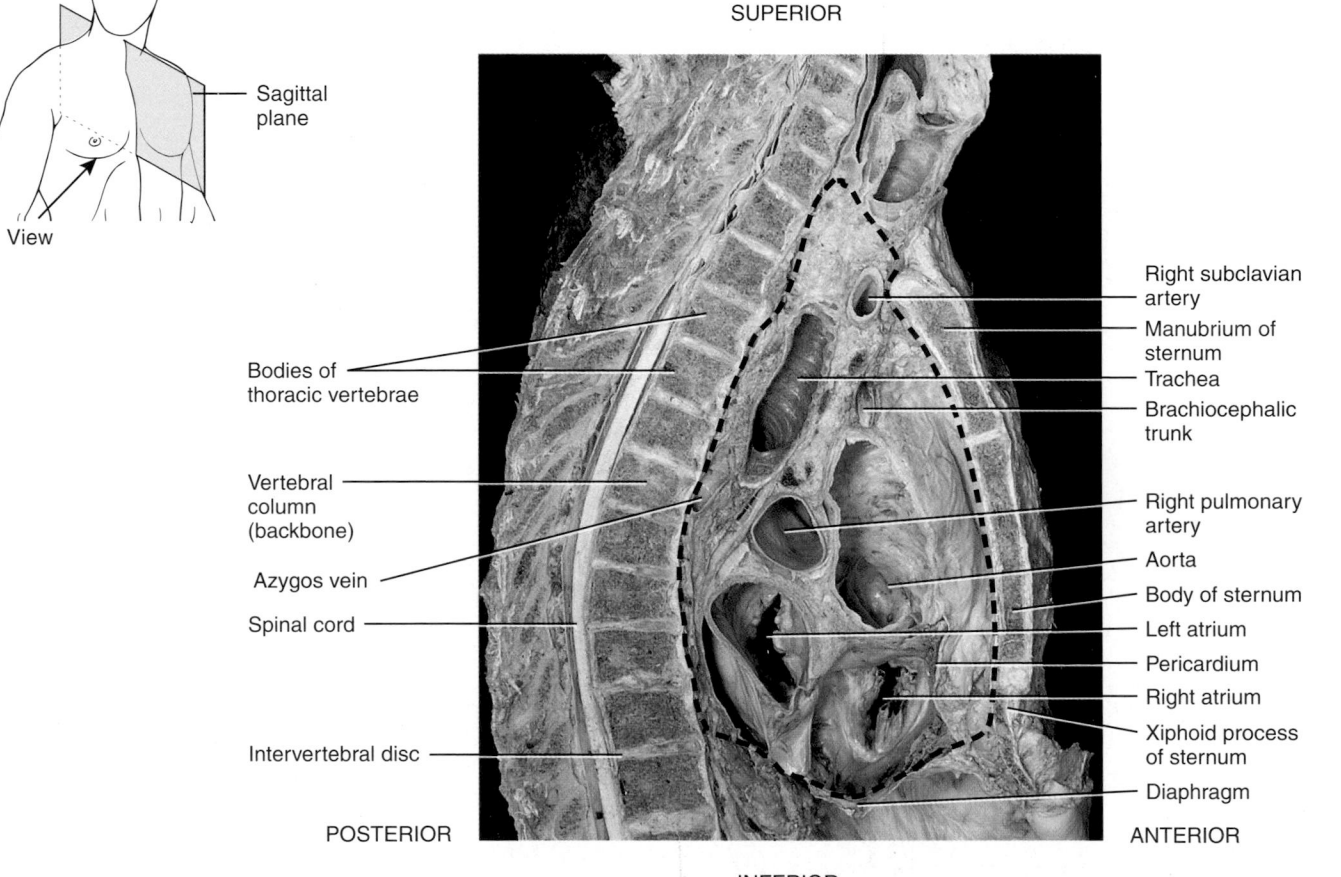

SUPERIOR

Sagittal plane

View

Bodies of thoracic vertebrae

Vertebral column (backbone)

Azygos vein

Spinal cord

Intervertebral disc

POSTERIOR

Right subclavian artery

Manubrium of sternum

Trachea

Brachiocephalic trunk

Right pulmonary artery

Aorta

Body of sternum

Left atrium

Pericardium

Right atrium

Xiphoid process of sternum

Diaphragm

ANTERIOR

INFERIOR

Sagittal section

FIGURE 6.5
Heart (the mediastinum is outlined by the dashed line)

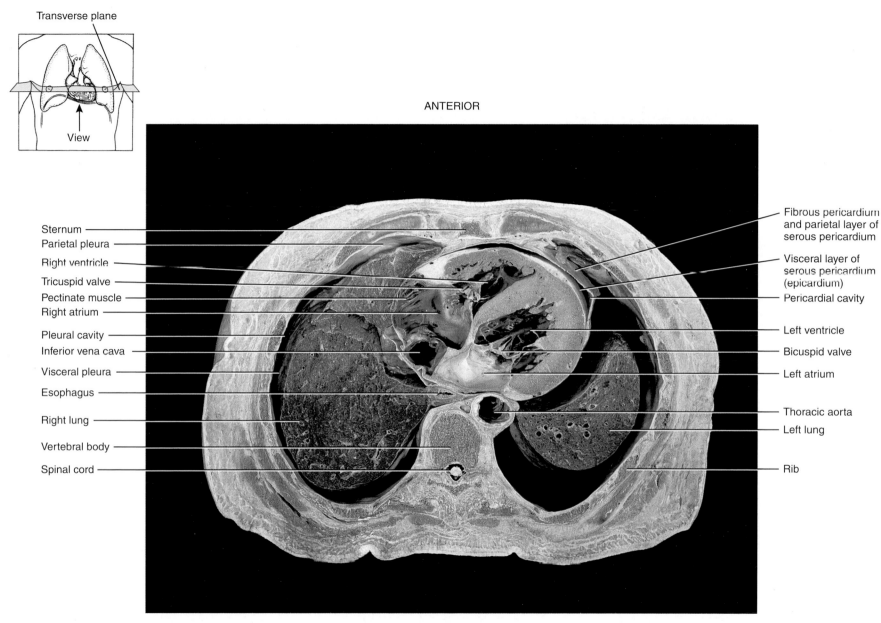

ANTERIOR

Sternum

Parietal pleura

Right ventricle

Tricuspid valve

Pectinate muscle

Right atrium

Pleural cavity

Inferior vena cava

Visceral pleura

Esophagus

Right lung

Vertebral body

Spinal cord

Fibrous pericardium and parietal layer of serous pericardium

Visceral layer of serous pericardium (epicardium)

Pericardial cavity

Left ventricle

Bicuspid valve

Left atrium

Thoracic aorta

Left lung

Rib

POSTERIOR

Inferior view of transverse section

Transverse plane

View

FIGURE 6.6
Heart

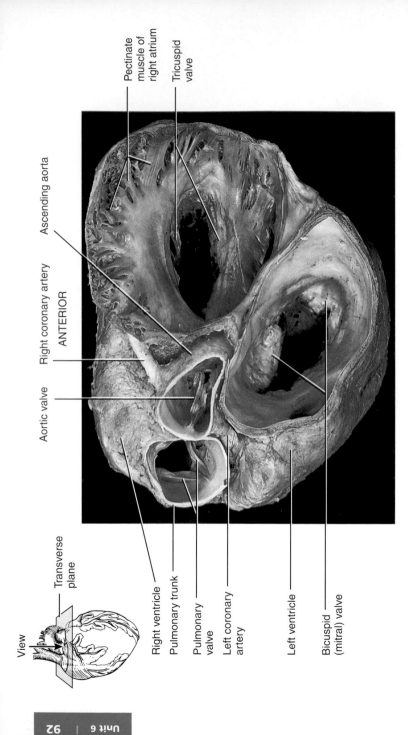

Pectinate
muscle of
right atrium

Tricuspid
valve

Ascending aorta

Right coronary artery

ANTERIOR

Aortic valve

POSTERIOR

Superior view of transverse section

View

Transverse
plane

Right ventricle

Pulmonary trunk

Pulmonary
valve

Left coronary
artery

Left ventricle

Bicuspid
(mitral) valve

F I G U R E   6 . 7
Heart valves

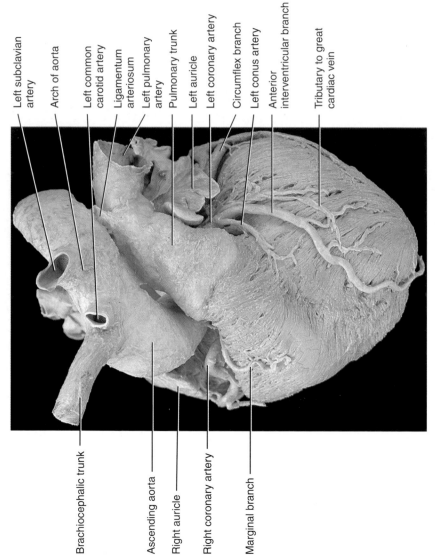

Left subclavian
artery

Arch of aorta

Left common
carotid artery

Ligamentum
arteriosum

Left pulmonary
artery

Pulmonary trunk

Left auricle

Left coronary artery

Circumflex branch

Left conus artery

Anterior
interventricular branch

Tributary to great
cardiac vein

SUPERIOR

Brachiocephalic trunk

Ascending aorta

Right auricle

Right coronary artery

Marginal branch

INFERIOR

Anterior view

F I G U R E   6 . 8
Blood supply to the heart

SUPERIOR

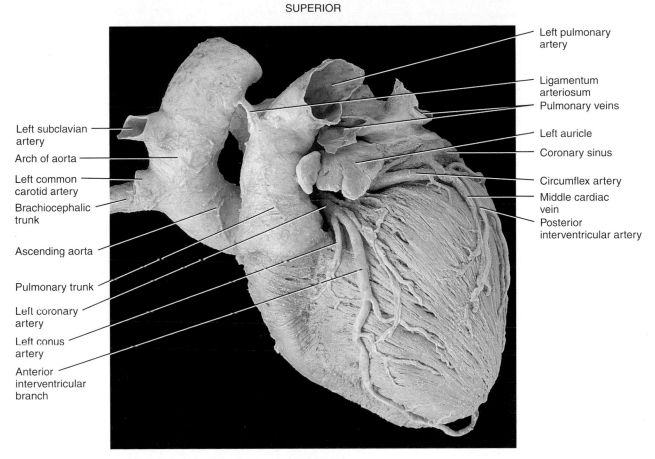

Left pulmonary
artery

Ligamentum
arteriosum

Pulmonary veins

Left auricle

Coronary sinus

Circumflex artery

Middle cardiac
vein

Posterior
interventricular artery

Left subclavian
artery

Arch of aorta

Left common
carotid artery

Brachiocephalic
trunk

Ascending aorta

Pulmonary trunk

Left coronary
artery

Left conus
artery

Anterior
interventricular
branch

INFERIOR

Anterolateral view

FIGURE 6.9
Blood supply to the heart

Right common carotid artery

Trachea

Brachiocephalic trunk

Ascending aorta

Esophagus

Inferior vena cava

Celiac trunk

Common hepatic artery

Superior mesenteric artery

Right common iliac artery

Right ureter

Larynx

Left common carotid artery

Left vertebral artery

Left subclavian artery

Arch of aorta

Left primary bronchus

Thoracic aorta

Diaphragm

Left adrenal (suprarenal) gland

Splenic artery

Left renal artery

Left kidney

Abdominal aorta

Inferior mesenteric artery

Psoas major muscle

Sigmoid colon

SUPERIOR

INFERIOR

Anterior view

F I G U R E   6 . 1 0

Arteries of the thorax, abdomen, and pelvis

SUPERIOR

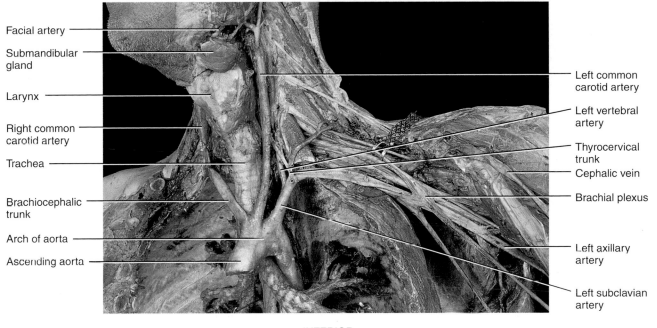

Facial artery

Submandibular
gland

Larynx

Right common
carotid artery

Trachea

Brachiocephalic
trunk

Arch of aorta

Ascending aorta

Left common
carotid artery

Left vertebral
artery

Thyrocervical
trunk

Cephalic vein

Brachial plexus

Left axillary
artery

Left subclavian
artery

INFERIOR

Anterior view

FIGURE 6.11
Arteries of the neck and shoulders

SUPERIOR

Medial pectoral
nerve

Pectoralis major
muscle

Pectoralis minor
muscle

Cephalic vein

Brachial artery

Ulnar nerve

Median nerve

Medial antebrachial
cutaneous nerve

Biceps brachii
muscle

Brachioradialis
muscle

Lateral antebrachial
cutaneous nerve

Transverse cervical
artery

Common carotid
artery

Subclavian artery

Suprascapular
artery

Clavicle (cut)

Axillary artery

Thoracoacromial artery

Lateral thoracic artery

Thoracodorsal nerve

Circumflex scapular
artery

Latissimus dorsi
muscle

Serratus anterior
muscle

External oblique
muscle

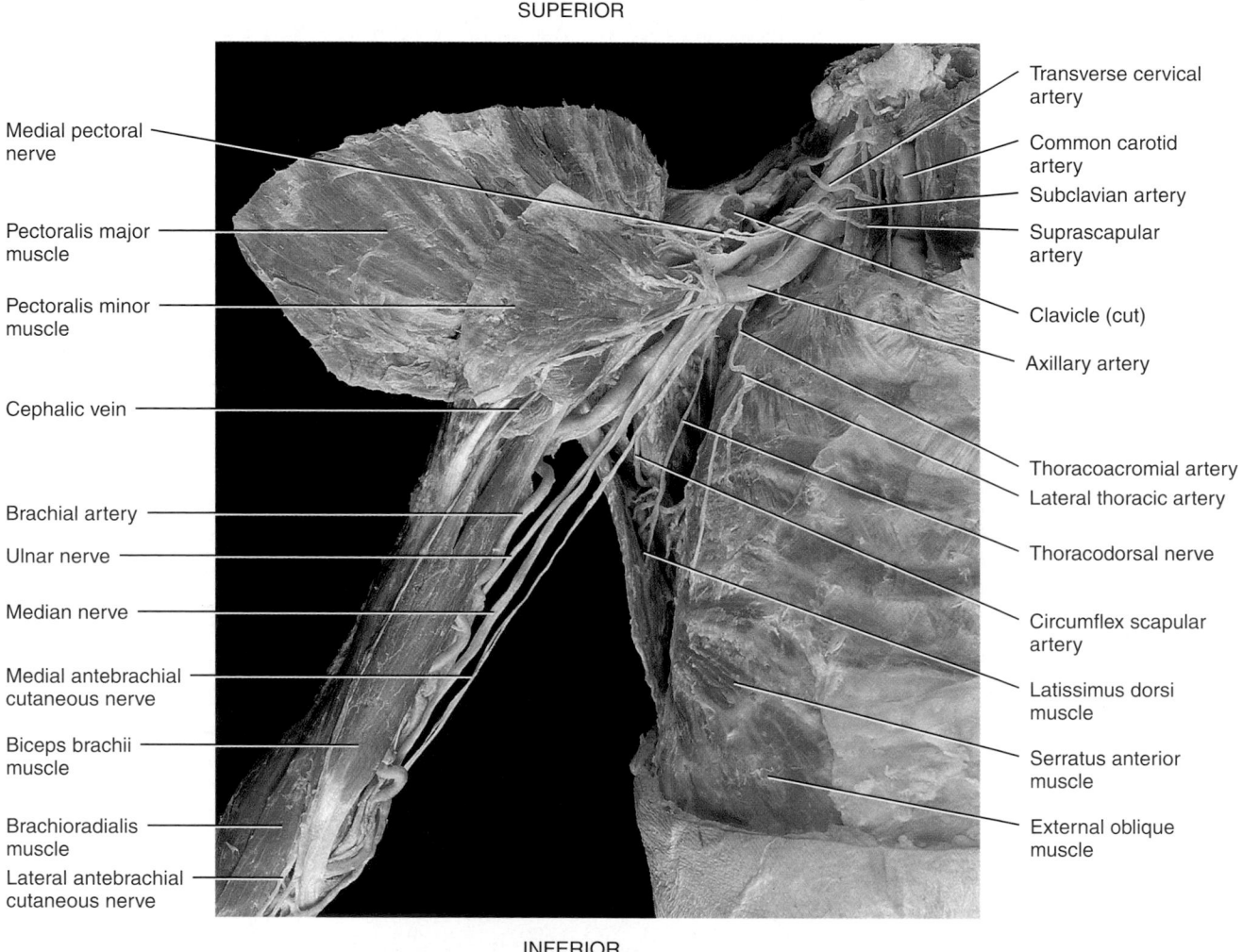

INFERIOR

Anterior view

FIGURE 6.12
*Arteries of the neck and
right upper limb*

SUPERIOR

Brachial artery

Anterior interosseous artery

Ulnar nerve

Median nerve

Radial artery

Ulnar artery

Hypothenar muscles

Thenar muscles

Superficial palmar arch

Common palmar digital artery

Proper palmar digital artery

LATERAL

MEDIAL

INFERIOR

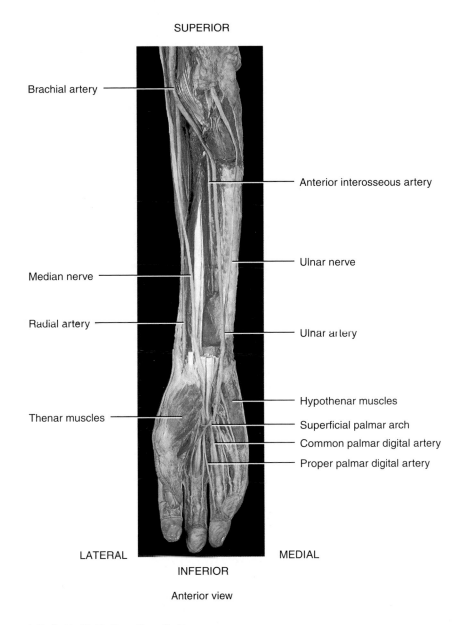

Anterior view

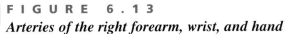

F I G U R E   6 . 1 3
Arteries of the right forearm, wrist, and hand

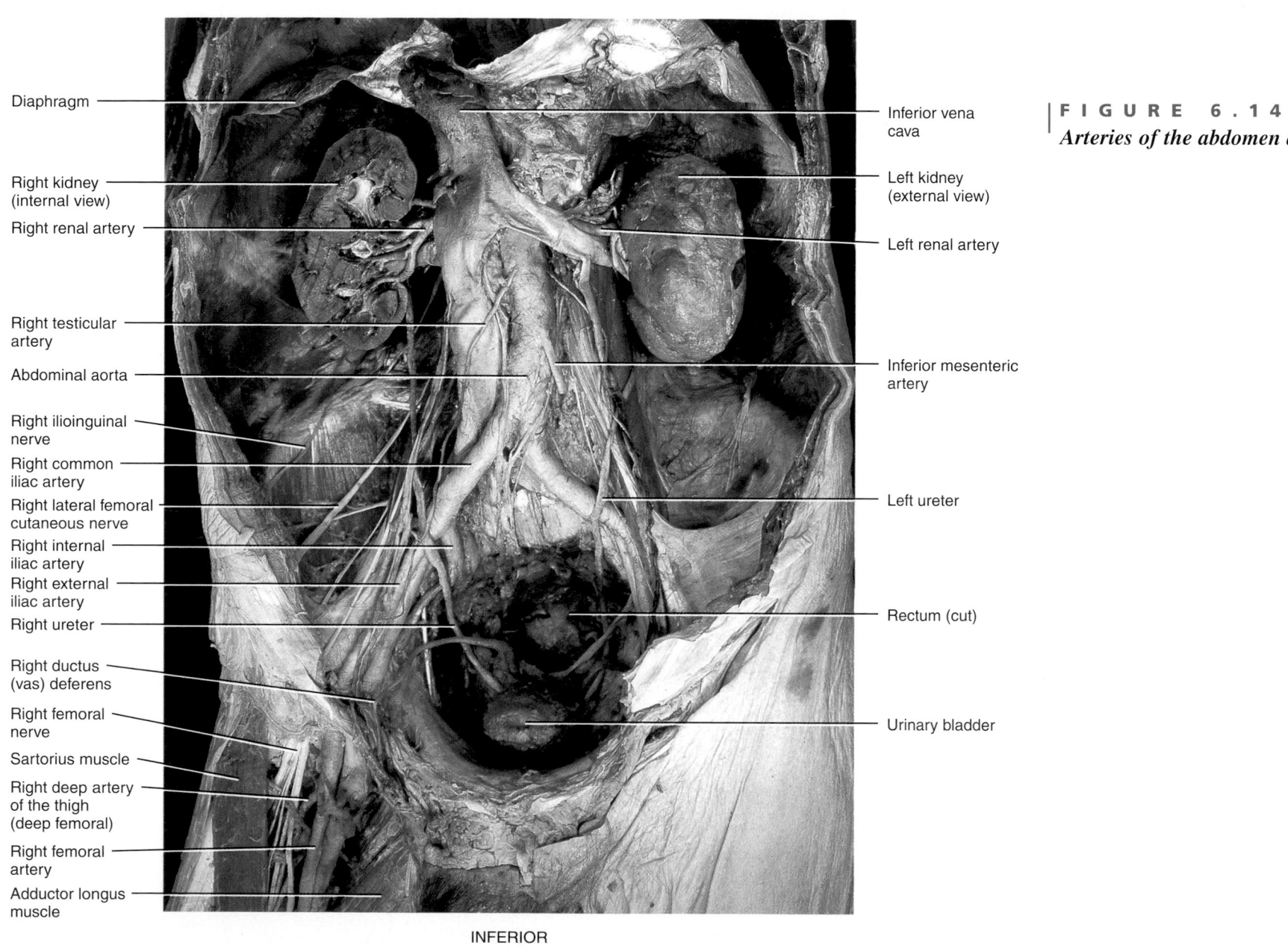

Diaphragm

Right kidney
(internal view)

Right renal artery

Right testicular
artery

Abdominal aorta

Right ilioinguinal
nerve

Right common
iliac artery

Right lateral femoral
cutaneous nerve

Right internal
iliac artery

Right external
iliac artery

Right ureter

Right ductus
(vas) deferens

Right femoral
nerve

Sartorius muscle

Right deep artery
of the thigh
(deep femoral)

Right femoral
artery

Adductor longus
muscle

Inferior vena
cava

Left kidney
(external view)

Left renal artery

Inferior mesenteric
artery

Left ureter

Rectum (cut)

Urinary bladder

INFERIOR

Anterior view

F I G U R E   6 . 1 4
Arteries of the abdomen and pelvis

SUPERIOR

Esophagus in
esophageal hiatus

Inferior phrenic
artery

Diaphragm

Inferior vena cava
(cut)

Celiac trunk

Left adrenal
(suprarenal) gland

Common hepatic
artery

Splenic artery

Left renal artery

Right renal vein
(cut)

Superior
mesenteric artery

Right ureter

Middle colic artery

Inferior
mesenteric artery

Right colic artery

Left colic artery

Abdominal aorta

Sigmoid artery

Ileocolic artery

Superior rectal
artery

Right common
iliac artery

Lateral femoral
cutaneous nerve

Sigmoid colon

Right external
iliac artery

Right external
iliac vein

INFERIOR

Anterior view

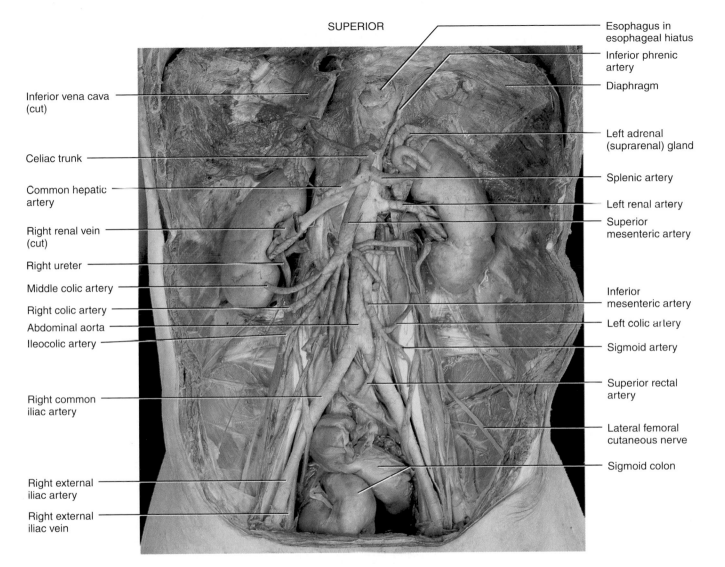

| FIGURE   6 . 1 5
Arteries of the abdomen and pelvis

SUPERIOR

Liver

Hepatic
portal vein

Superior
mesenteric vein

Right colic vein

Ileocolic vein

Ascending colon

Transverse
colon

Descending
colon

Splenic vein
(cut)

Superior
mesenteric
artery

Jejunum

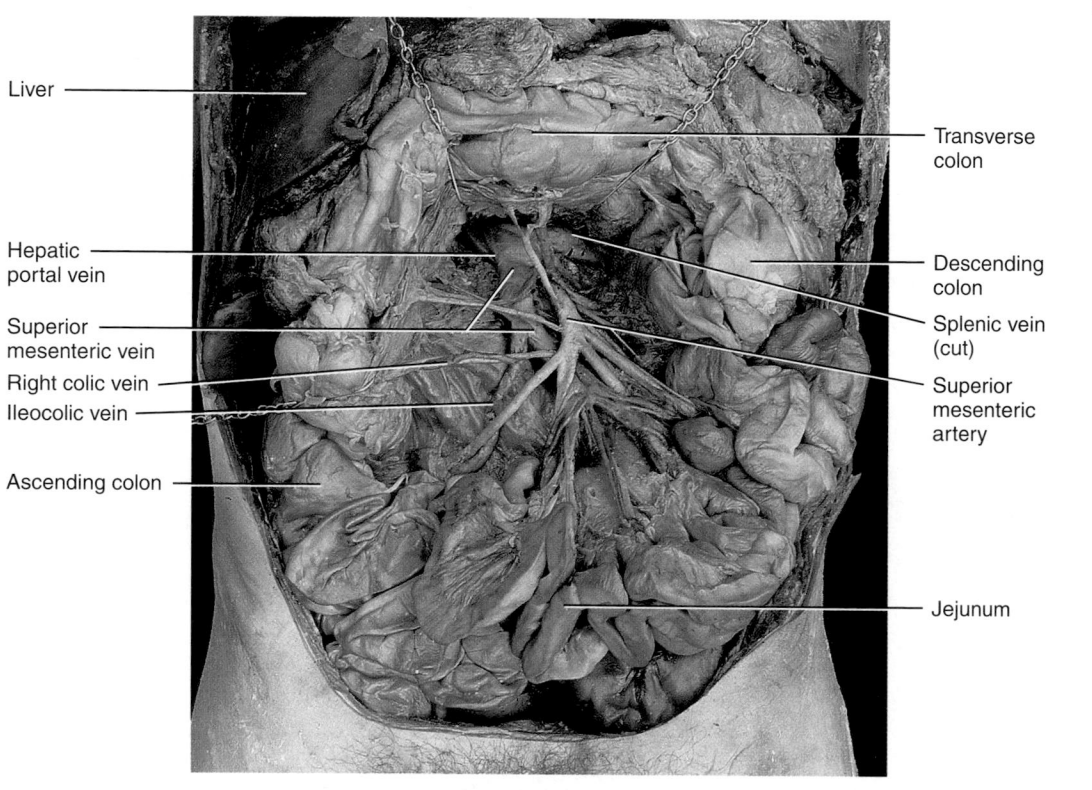

INFERIOR

Anterior view

FIGURE 6.16
Tributaries of the hepatic portal vein and branches of the superior mesenteric artery

SUPERIOR

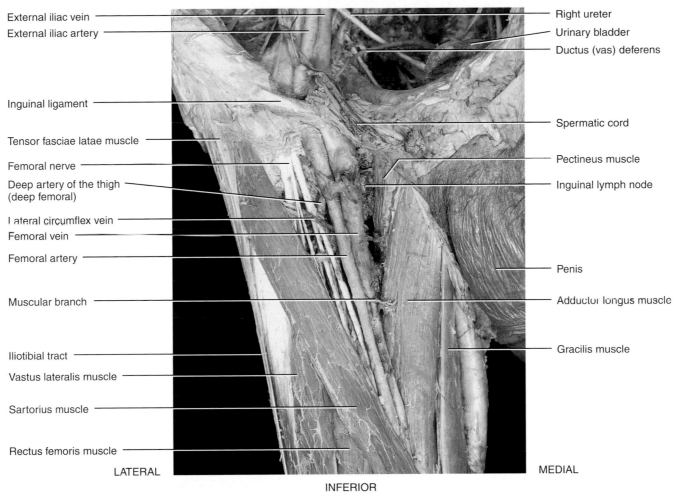

External iliac vein

External iliac artery

Inguinal ligament

Tensor fasciae latae muscle

Femoral nerve

Deep artery of the thigh
(deep femoral)

Lateral circumflex vein

Femoral vein

Femoral artery

Muscular branch

Iliotibial tract

Vastus lateralis muscle

Sartorius muscle

Rectus femoris muscle

Right ureter

Urinary bladder

Ductus (vas) deferens

Spermatic cord

Pectineus muscle

Inguinal lymph node

Penis

Adductor longus muscle

Gracilis muscle

LATERAL

INFERIOR

MEDIAL

Anterior view

FIGURE 6.17
Blood vessels of the right thigh

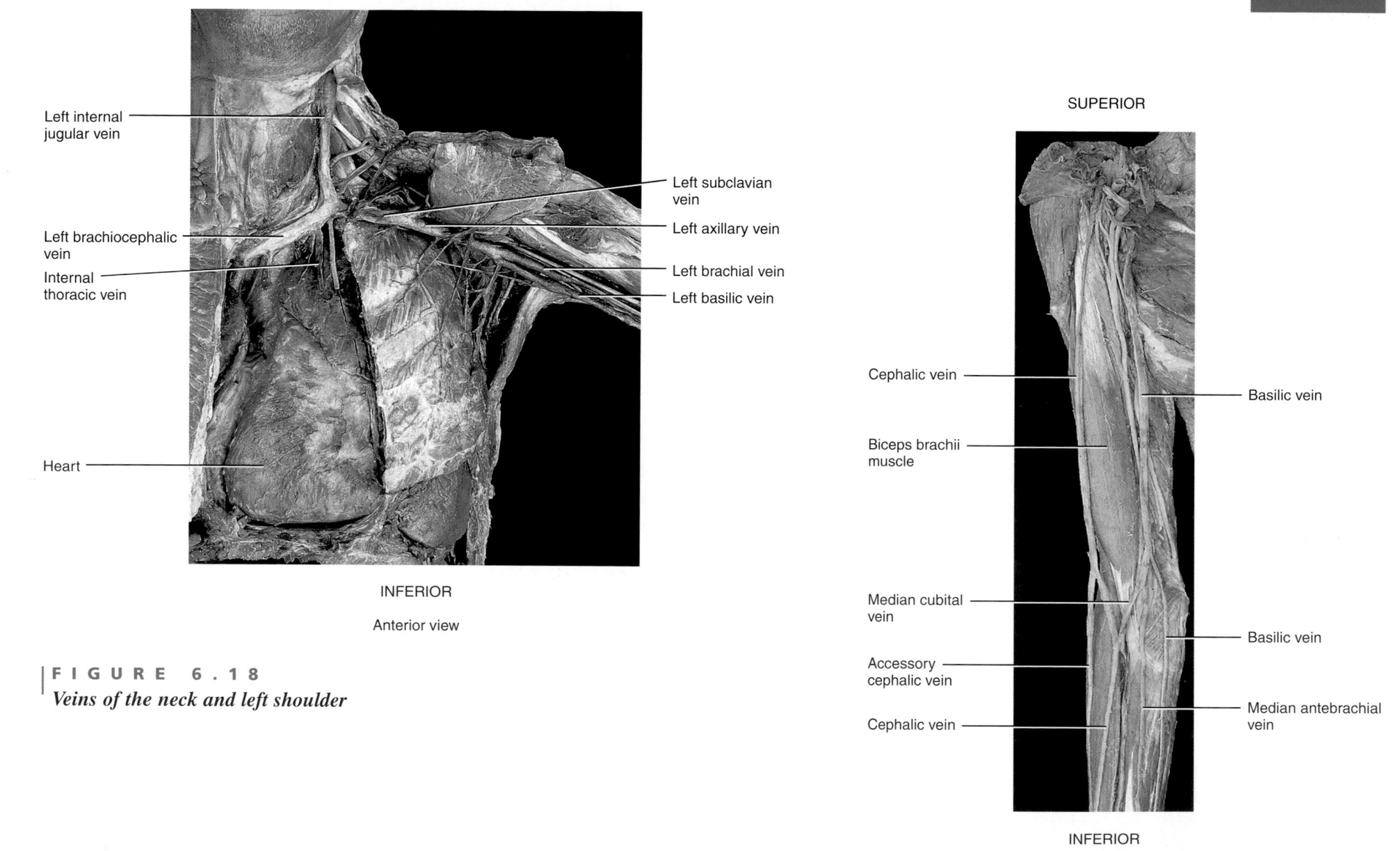

Left internal jugular vein

Left brachiocephalic vein

Internal thoracic vein

Heart

Left subclavian vein

Left axillary vein

Left brachial vein

Left basilic vein

INFERIOR

Anterior view

FIGURE 6.18

Veins of the neck and left shoulder

SUPERIOR

Cephalic vein

Biceps brachii muscle

Median cubital vein

Accessory cephalic vein

Cephalic vein

Basilic vein

Basilic vein

Median antebrachial vein

INFERIOR

Anterior view

FIGURE 6.19

Veins of the right arm and forearm

SUPERIOR

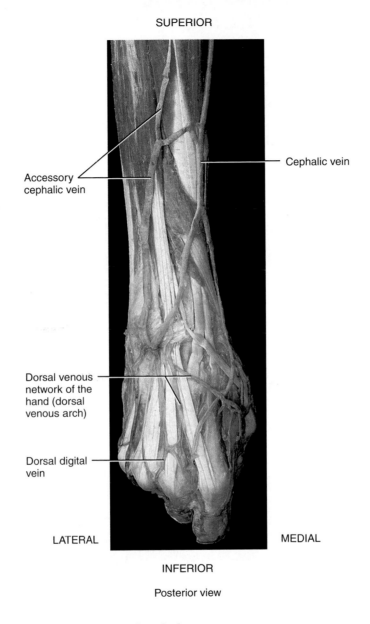

Accessory
cephalic vein

Cephalic vein

Dorsal venous
network of the
hand (dorsal
venous arch)

Dorsal digital
vein

LATERAL

MEDIAL

INFERIOR

Posterior view

FIGURE 6.20
Veins of the right forearm and hand

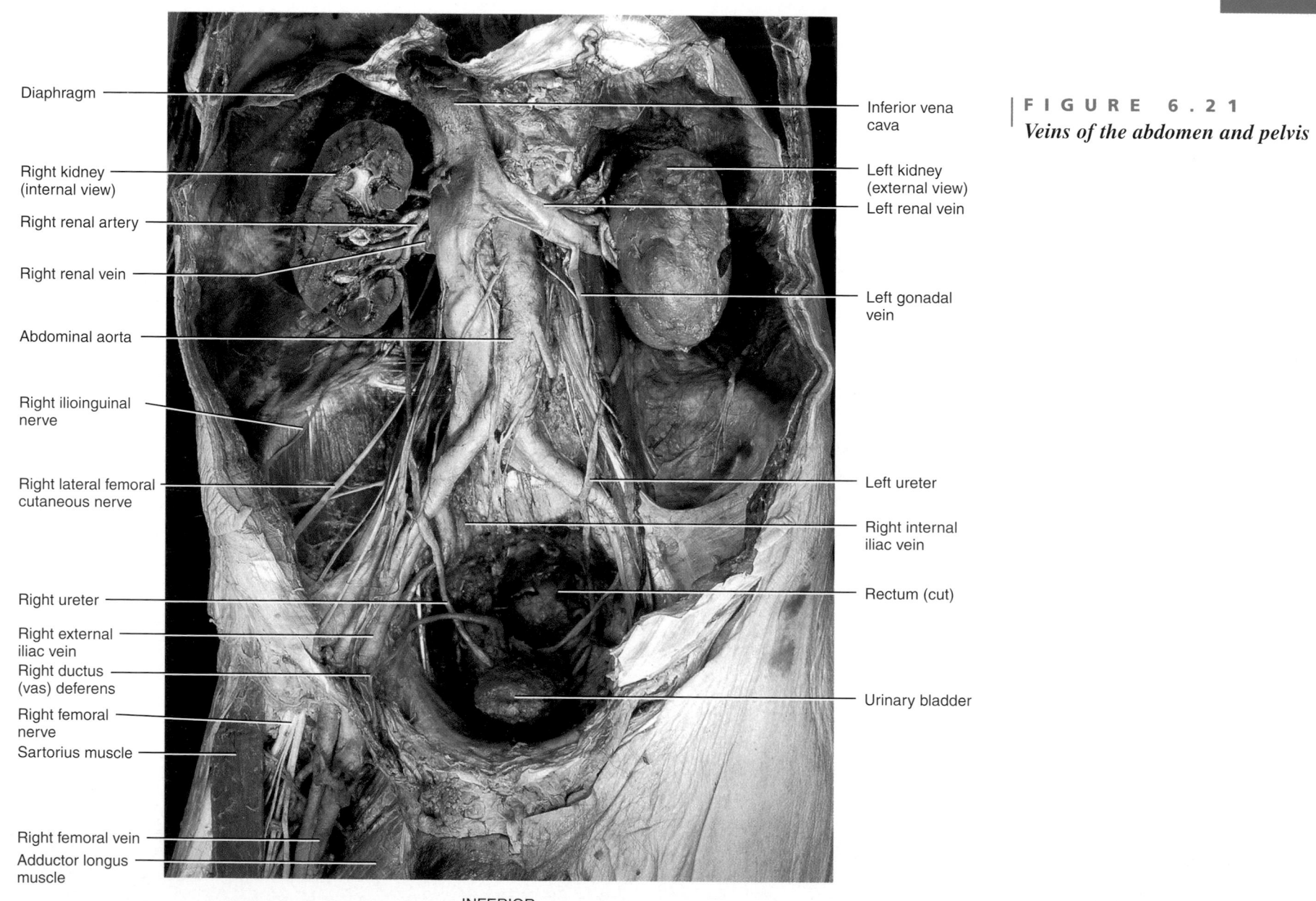

Diaphragm

Right kidney
(internal view)

Right renal artery

Right renal vein

Abdominal aorta

Right ilioinguinal
nerve

Right lateral femoral
cutaneous nerve

Right ureter

Right external
iliac vein

Right ductus
(vas) deferens

Right femoral
nerve

Sartorius muscle

Right femoral vein

Adductor longus
muscle

Inferior vena
cava

Left kidney
(external view)

Left renal vein

Left gonadal
vein

Left ureter

Right internal
iliac vein

Rectum (cut)

Urinary bladder

FIGURE 6.21
Veins of the abdomen and pelvis

INFERIOR

Anterior view

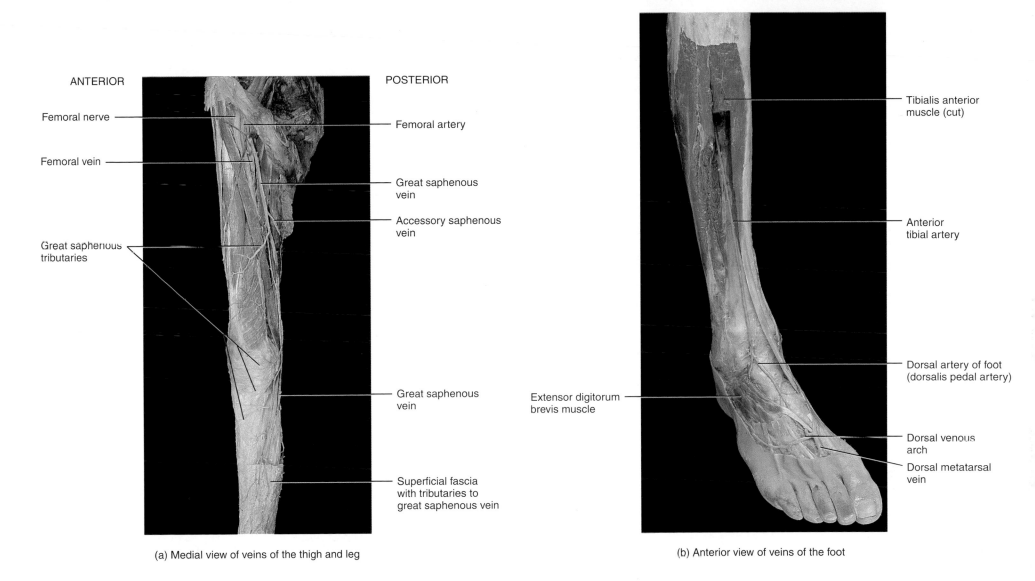

ANTERIOR

POSTERIOR

Femoral nerve

Femoral vein

Great saphenous
tributaries

Femoral artery

Great saphenous
vein

Accessory saphenous
vein

Great saphenous
vein

Superficial fascia
with tributaries to
great saphenous vein

(a) Medial view of veins of the thigh and leg

Tibialis anterior
muscle (cut)

Anterior
tibial artery

Dorsal artery of foot
(dorsalis pedal artery)

Extensor digitorum
brevis muscle

Dorsal venous
arch

Dorsal metatarsal
vein

(b) Anterior view of veins of the foot

FIGURE 6.22
Veins of the right lower limb

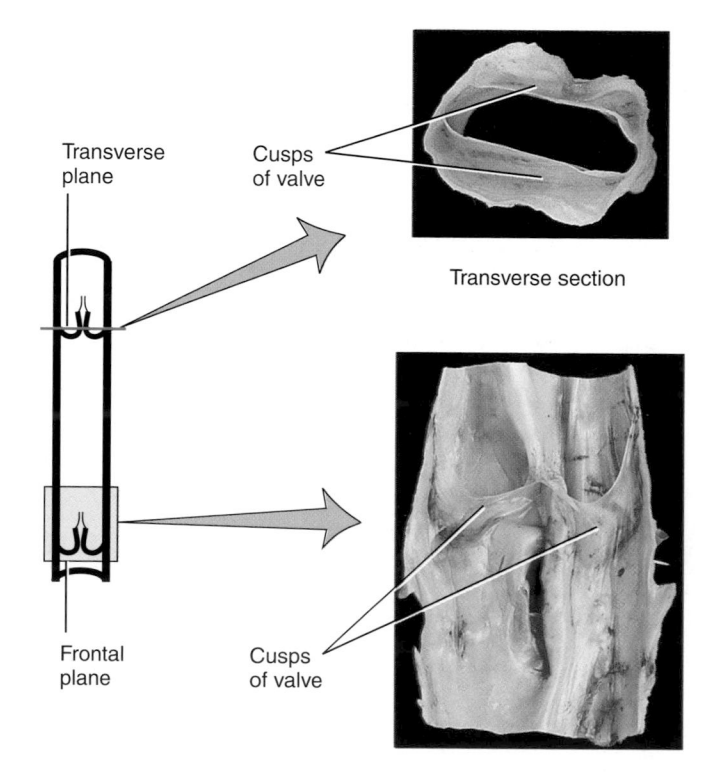

Transverse
plane

Cusps
of valve

Transverse section

Frontal
plane

Cusps
of valve

Longitudinally cut

FIGURE 6.23
Valves in a vein

UNIT SEVEN | *The Lymphatic and Immune System*

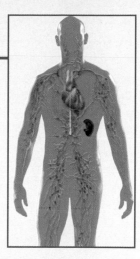

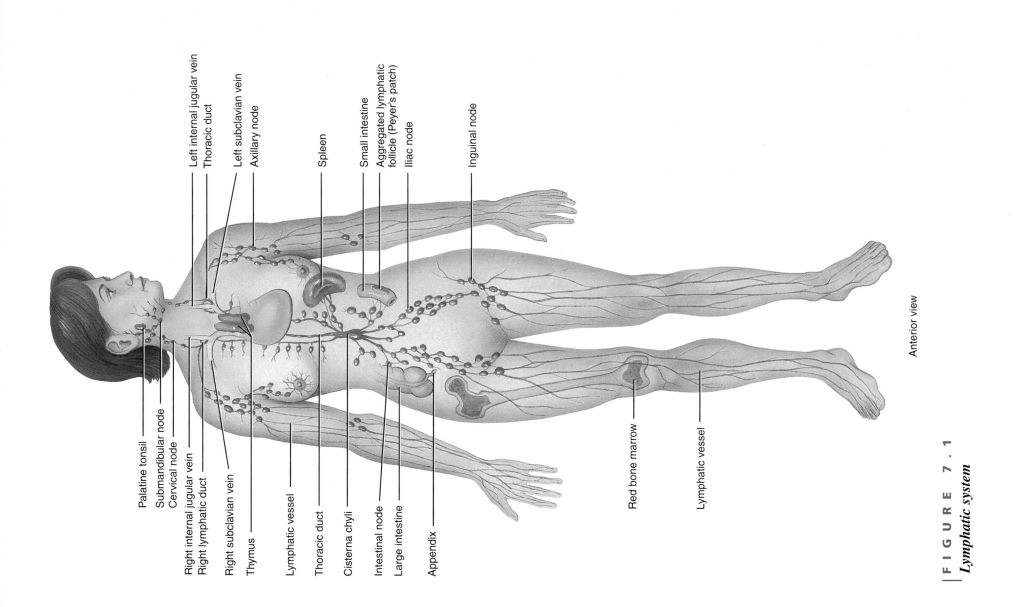

Palatine tonsil
Submandibular node
Cervical node
Right internal jugular vein
Right lymphatic duct
Right subclavian vein
Thymus
Lymphatic vessel
Thoracic duct
Cisterna chyli
Intestinal node
Large intestine
Appendix

Left internal jugular vein
Thoracic duct
Left subclavian vein
Axillary node
Spleen
Small intestine
Aggregated lymphatic follicle (Peyer's patch)
Iliac node
Inguinal node

Red bone marrow
Lymphatic vessel

Anterior view

F I G U R E   7 . 1
Lymphatic system

SUPERIOR

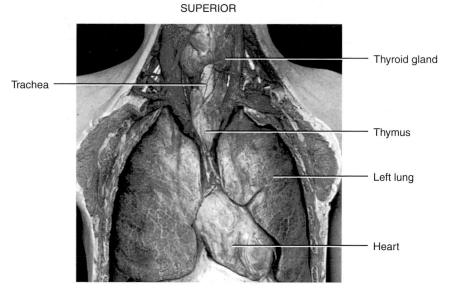

Thyroid gland

Trachea

Thymus

Left lung

Heart

INFERIOR

(a) Anterior view of thymus

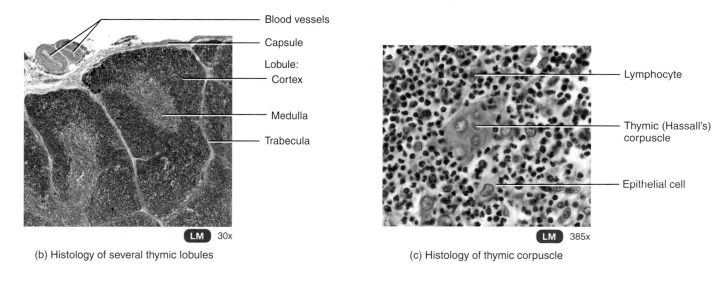

Blood vessels

Capsule

Lobule:
Cortex

Medulla

Trabecula

LM 30x

(b) Histology of several thymic lobules

Lymphocyte

Thymic (Hassall's)
corpuscle

Epithelial cell

LM 385x

(c) Histology of thymic corpuscle

FIGURE 7.2
Thymus

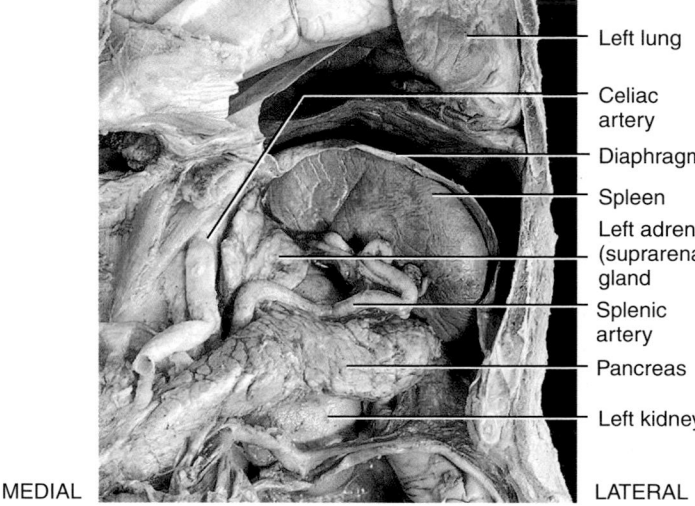
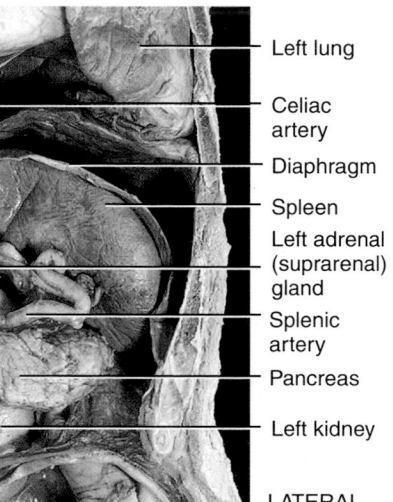

SUPERIOR

Left lung

Celiac artery

Diaphragm

Spleen

Left adrenal (suprarenal) gland

Splenic artery

Pancreas

Left kidney

MEDIAL

LATERAL

INFERIOR

(a) Anterior view of the spleen

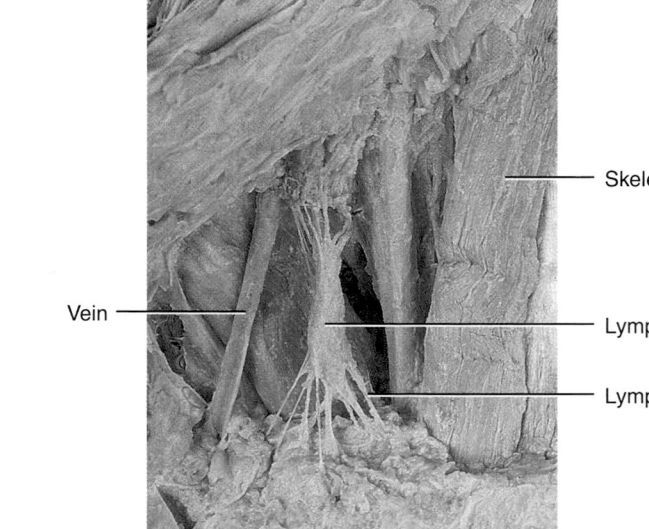

FIGURE 7.4
Lymph node

Skeletal muscle

Vein

Lymph node

Lymphatic vessel

(a) Anterior view of a lymph node

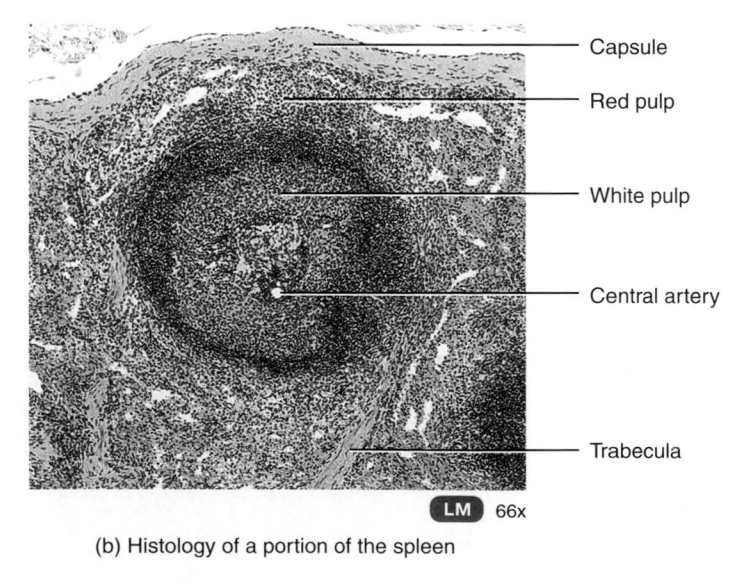

Capsule

Red pulp

White pulp

Central artery

Trabecula

LM 66x

(b) Histology of a portion of the spleen

FIGURE 7.3 *Spleen*

Capsule
Subcapsular sinus
Trabecula
Trabecular sinus
Outer cortex

Germinal center in secondary lymphatic nodule

Inner cortex

Medullary sinus

Medulla

LM 55x

(b) Histology of a portion of a lymph node

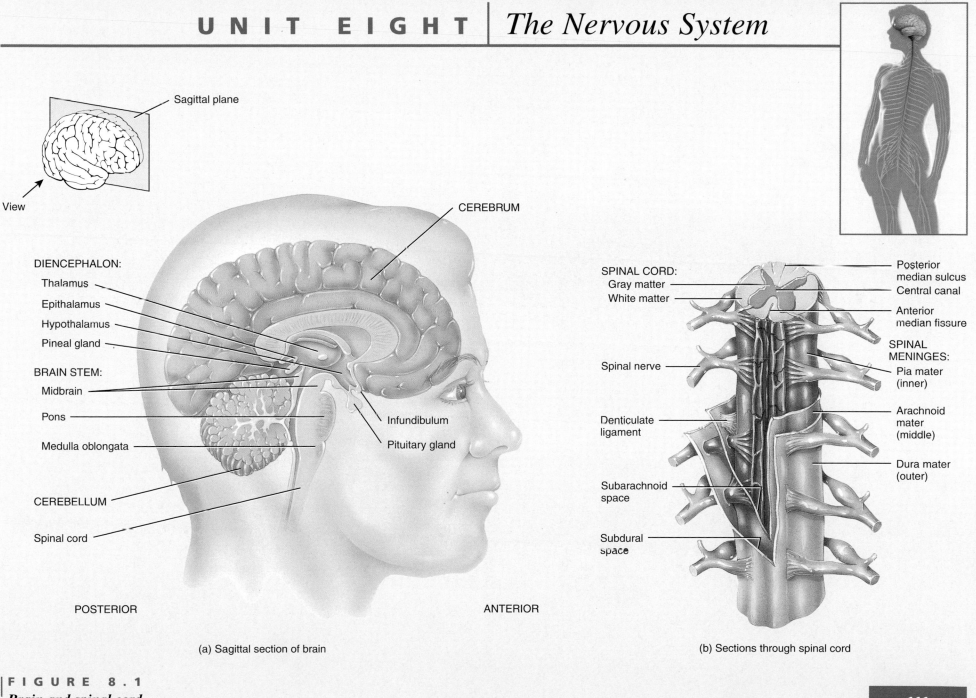

Sagittal plane

View

CEREBRUM

DIENCEPHALON:

Thalamus

Epithalamus

Hypothalamus

Pineal gland

BRAIN STEM:

Midbrain

Pons

Medulla oblongata

CEREBELLUM

Spinal cord

Infundibulum

Pituitary gland

POSTERIOR

ANTERIOR

(a) Sagittal section of brain

SPINAL CORD:

Gray matter

White matter

Posterior median sulcus

Central canal

Anterior median fissure

SPINAL MENINGES:

Spinal nerve

Pia mater (inner)

Denticulate ligament

Arachnoid mater (middle)

Subarachnoid space

Dura mater (outer)

Subdural space

(b) Sections through spinal cord

FIGURE 8.1
Brain and spinal cord

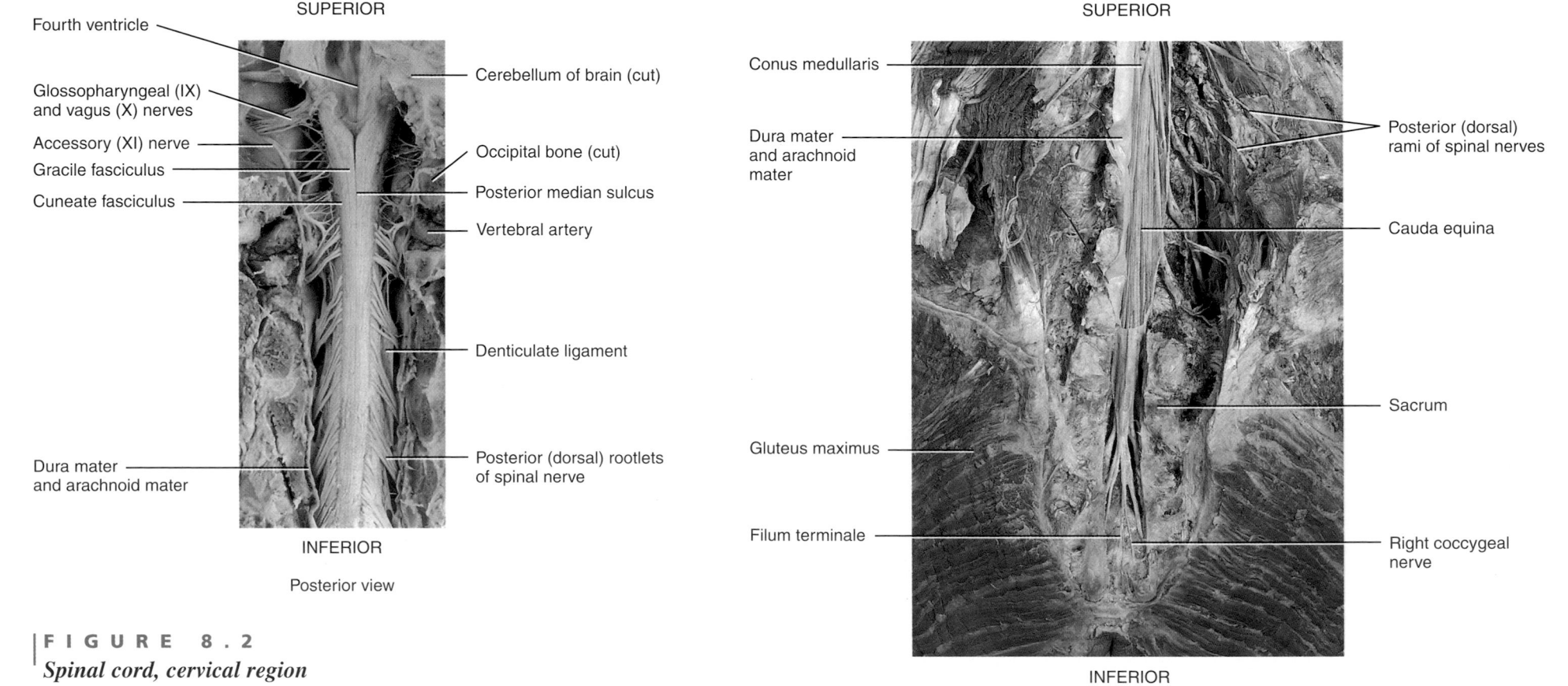

SUPERIOR

Fourth ventricle

Cerebellum of brain (cut)

Glossopharyngeal (IX) and vagus (X) nerves

Accessory (XI) nerve

Occipital bone (cut)

Gracile fasciculus

Posterior median sulcus

Cuneate fasciculus

Vertebral artery

Denticulate ligament

Dura mater and arachnoid mater

Posterior (dorsal) rootlets of spinal nerve

INFERIOR

Posterior view

FIGURE 8.2
Spinal cord, cervical region

SUPERIOR

Conus medullaris

Dura mater and arachnoid mater

Posterior (dorsal) rami of spinal nerves

Cauda equina

Sacrum

Gluteus maximus

Filum terminale

Right coccygeal nerve

INFERIOR

Posterior view

FIGURE 8.3
Spinal cord, lumbosacral region

POSTERIOR

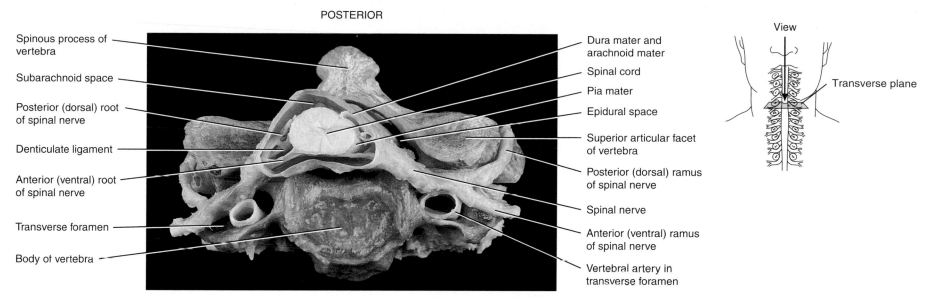

Spinous process of vertebra

Subarachnoid space

Posterior (dorsal) root of spinal nerve

Denticulate ligament

Anterior (ventral) root of spinal nerve

Transverse foramen

Body of vertebra

Dura mater and arachnoid mater

Spinal cord

Pia mater

Epidural space

Superior articular facet of vertebra

Posterior (dorsal) ramus of spinal nerve

Spinal nerve

Anterior (ventral) ramus of spinal nerve

Vertebral artery in transverse foramen

ANTERIOR

Superior view of transverse section

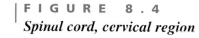

View

Transverse plane

FIGURE 8.4
Spinal cord, cervical region

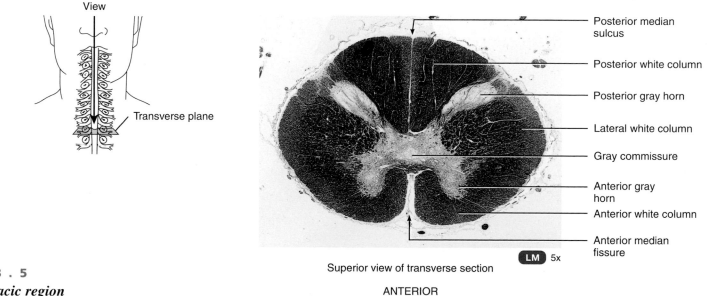

View

Transverse plane

Posterior median sulcus

Posterior white column

Posterior gray horn

Lateral white column

Gray commissure

Anterior gray horn

Anterior white column

Anterior median fissure

LM 5x

Superior view of transverse section

ANTERIOR

FIGURE 8.5
Spinal cord, thoracic region

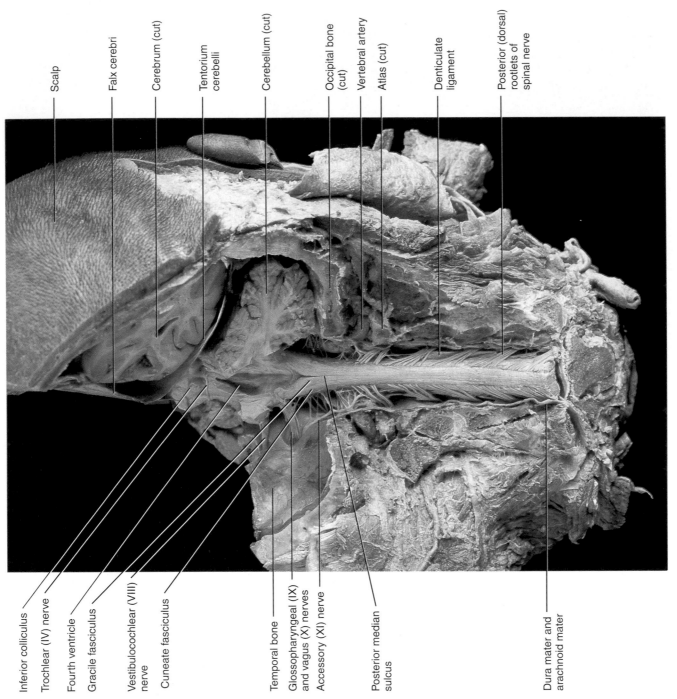

SUPERIOR

INFERIOR

Scalp

Falx cerebri

Cerebrum (cut)

Tentorium cerebelli

Cerebellum (cut)

Occipital bone (cut)

Vertebral artery

Atlas (cut)

Denticulate ligament

Posterior (dorsal) rootlets of spinal nerve

Inferior colliculus

Trochlear (IV) nerve

Fourth ventricle

Gracile fasciculus

Vestibulocochlear (VIII) nerve

Cuneate fasciculus

Temporal bone

Glossopharyngeal (IX) and vagus (X) nerves

Accessory (XI) nerve

Posterior median sulcus

Dura mater and arachnoid mater

F I G U R E   8 . 6   *Spinal cord (posterior view) and brain (oblique section)*

SUPERIOR

POSTERIOR

ANTERIOR

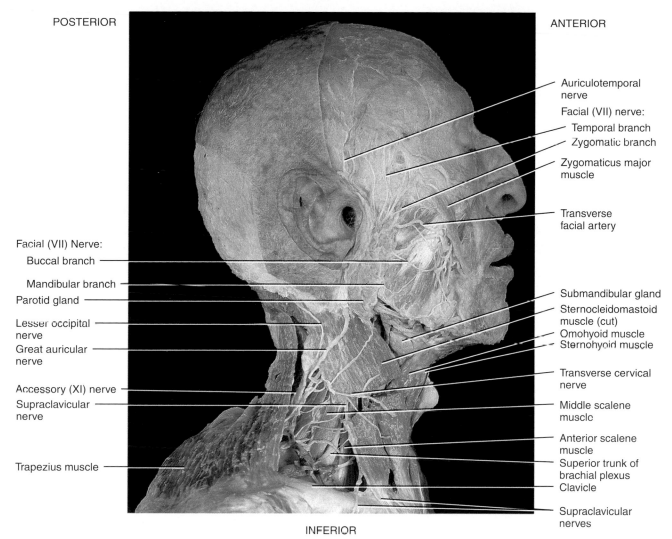

Auriculotemporal
nerve

Facial (VII) nerve:

Temporal branch

Zygomatic branch

Zygomaticus major
muscle

Transverse
facial artery

Facial (VII) Nerve:

Buccal branch

Mandibular branch

Parotid gland

Lesser occipital
nerve

Great auricular
nerve

Accessory (XI) nerve

Supraclavicular
nerve

Trapezius muscle

Submandibular gland

Sternocleidomastoid
muscle (cut)

Omohyoid muscle

Sternohyoid muscle

Transverse cervical
nerve

Middle scalene
muscle

Anterior scalene
muscle

Superior trunk of
brachial plexus

Clavicle

Supraclavicular
nerves

INFERIOR

Right lateral view

FIGURE 8.7
Facial nerves and cervical plexus

SUPERIOR

Medial pectoral nerve

Trunks of brachial plexus

Anterior scalene muscle

Phrenic nerve

Subclavian artery

Pectoralis minor muscle

Lateral pectoral nerve

Musculocutaneous nerve

Ulnar nerve

Medial cutaneous
nerve of forearm

Intercostobrachial nerves

Axillary artery

Phrenic nerve

Long thoracic nerve

Intercostal nerves

Median nerve

Rib (cut)

Inferior vena cava

Diaphragm

Pericardium (cut)

LATERAL

MEDIAL

INFERIOR

Anterior view

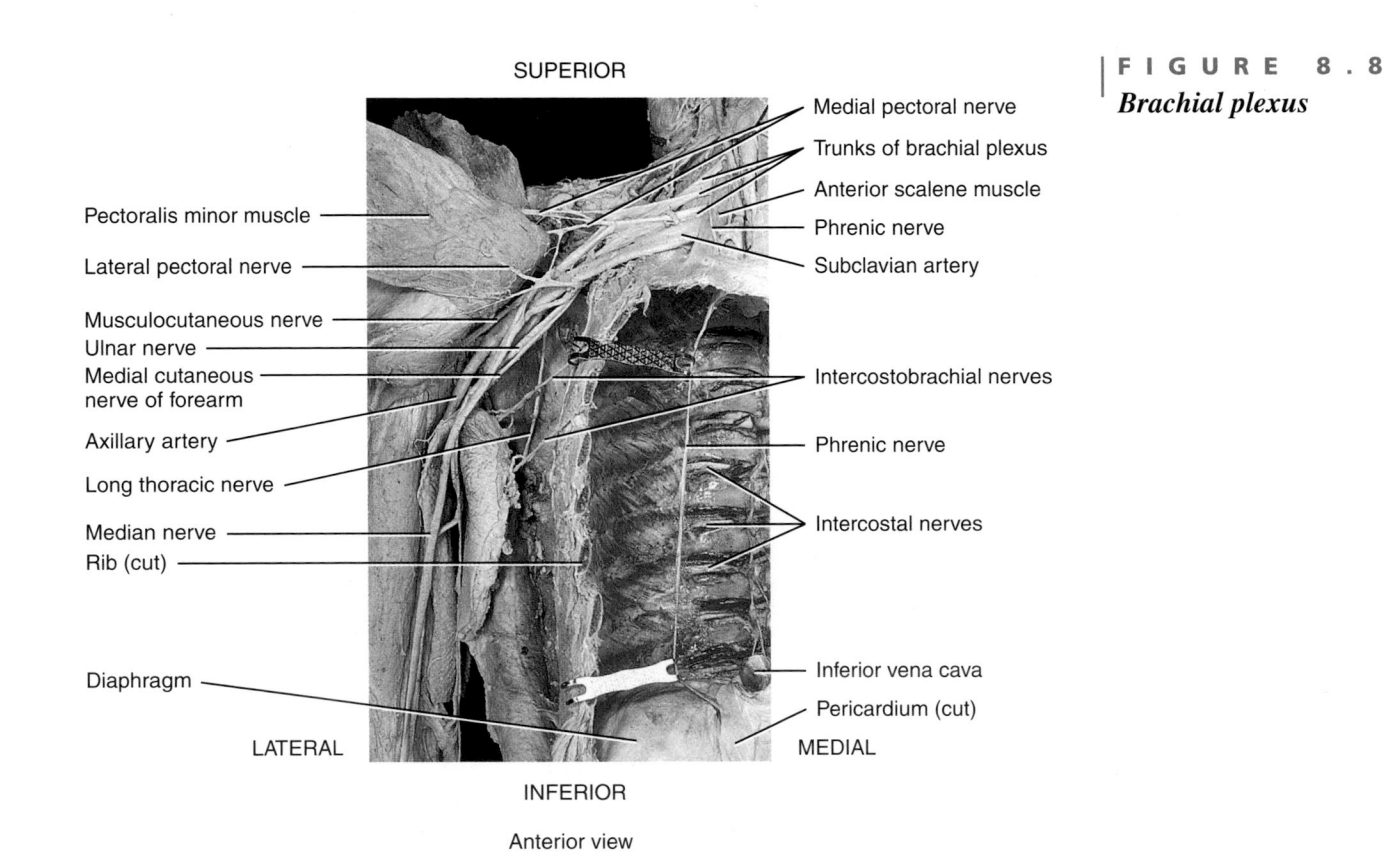

FIGURE 8.8
Brachial plexus

SUPERIOR

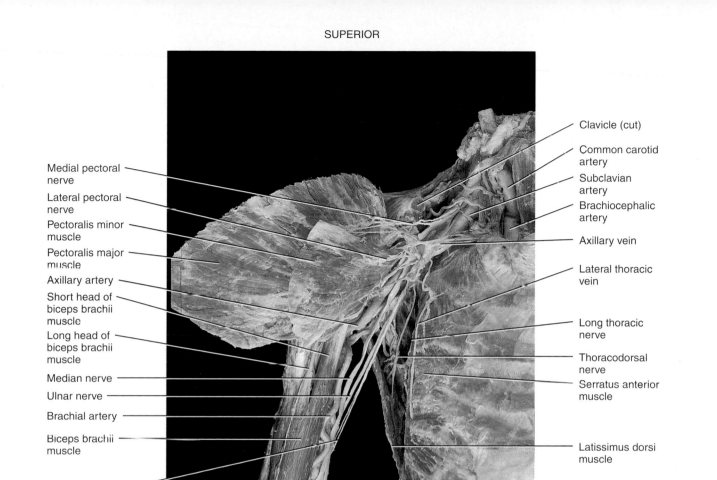

Clavicle (cut)

Common carotid
artery

Subclavian
artery

Brachiocephalic
artery

Axillary vein

Lateral thoracic
vein

Long thoracic
nerve

Thoracodorsal
nerve

Serratus anterior
muscle

Latissimus dorsi
muscle

Medial pectoral
nerve

Lateral pectoral
nerve

Pectoralis minor
muscle

Pectoralis major
muscle

Axillary artery

Short head of
biceps brachii
muscle

Long head of
biceps brachii
muscle

Median nerve

Ulnar nerve

Brachial artery

Biceps brachii
muscle

Medial cutaneous
nerve of forearm

LATERAL

MEDIAL

INFERIOR

Anterior view

FIGURE 8.9
Brachial plexus

SUPERIOR

Testicular artery

Ureter

Ilioinguinal nerve
Lateral femoral
cutaneous nerve
Iliacus muscle
Genitofemoral nerve
(femoral branch)

External iliac artery

Internal iliac artery

Genitofemoral nerve
(genital branch)
Ductus (vas) deferens

Inguinal ligament
Obturator nerve
Femoral nerve

Urinary bladder
Femoral artery
Femoral vein

Kidney
Abdominal aorta
Testicular vein
Inferior mesenteric
artery
Genitofemoral nerve
(genital branch)

Common iliac artery
Nerve to iliacus
Ureter

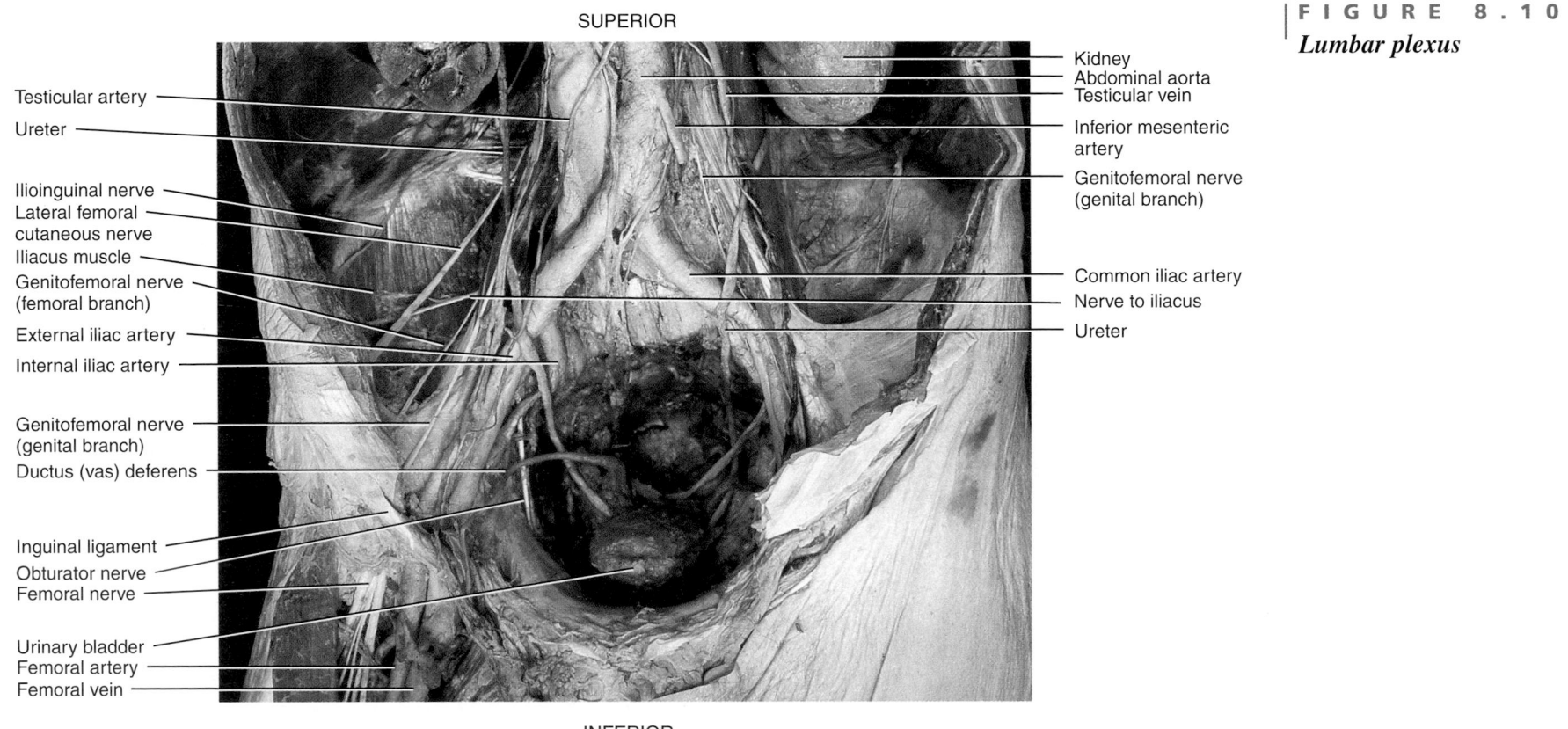

INFERIOR

Anterior view

| FIGURE 8.10
Lumbar plexus

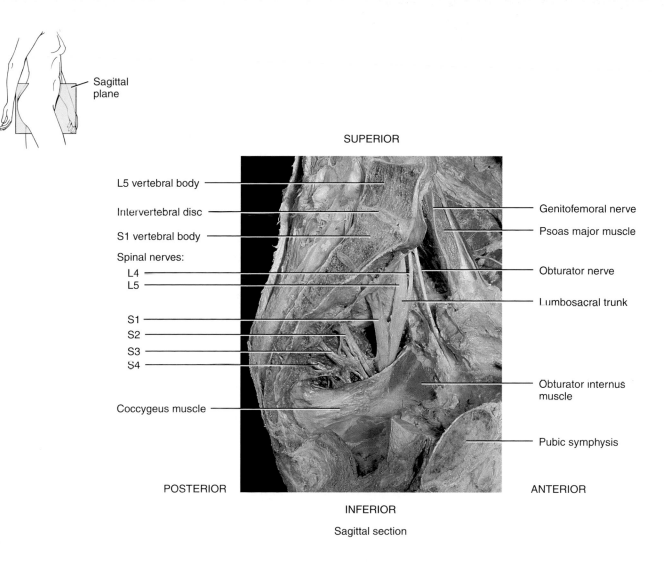

Sagittal plane

SUPERIOR

L5 vertebral body

Intervertebral disc

S1 vertebral body

Spinal nerves:
  L4
  L5
  S1
  S2
  S3
  S4

Coccygeus muscle

Genitofemoral nerve

Psoas major muscle

Obturator nerve

Lumbosacral trunk

Obturator internus muscle

Pubic symphysis

POSTERIOR

ANTERIOR

INFERIOR

Sagittal section

FIGURE 8.11
Sacral plexus

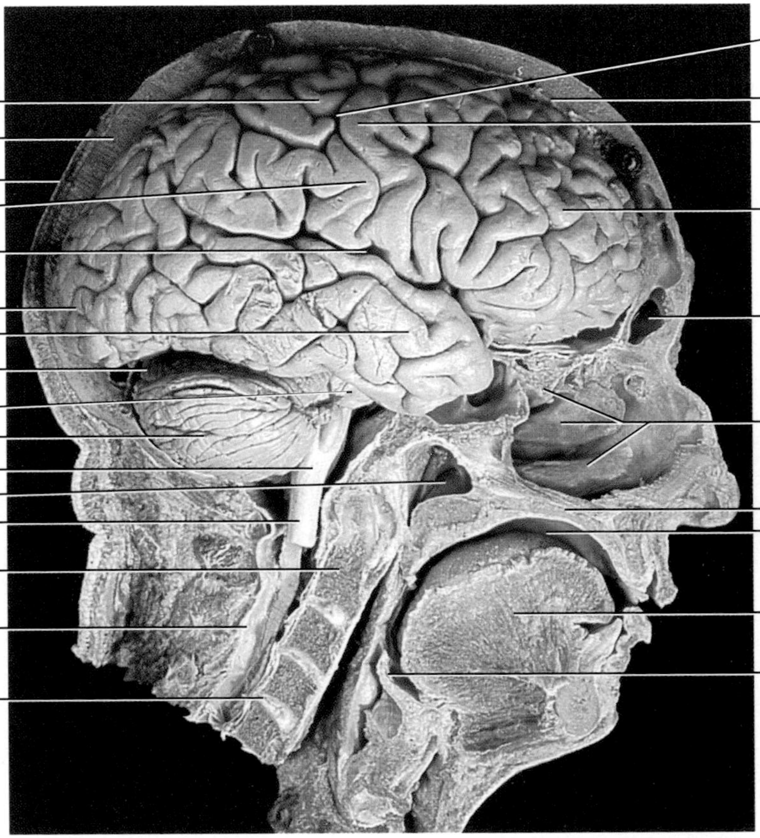
SUPERIOR
INFERIOR
POSTERIOR
ANTERIOR
Postcentral gyrus
Parietal bone (cut)
Skin
Parietal lobe
Lateral cerebral sulcus
Occipital lobe
Temporal lobe
Transverse fissure
Pons
Cerebellum
Medulla oblongata
Nasopharynx
Spinal cord
Body of cervical vertebra
Dura mater of spinal cord
Intervertebral disc
Central sulcus
Dura mater of brain
Precentral gyrus
Frontal lobe
Frontal sinus
Nasal conchae (turbinates)
Hard palate
Oral cavity
Tongue
Epiglottis

Right lateral view

FIGURE 8.12
Brain in cranial cavity

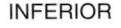

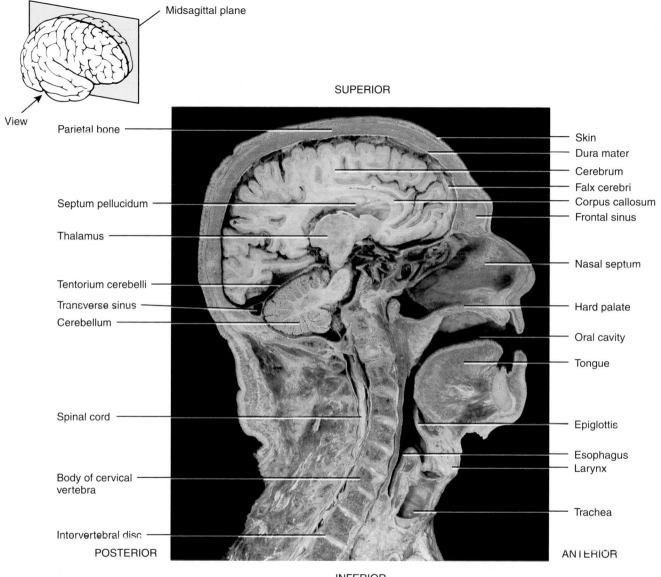

Midsagittal plane

View

SUPERIOR

Parietal bone

Skin

Dura mater

Cerebrum

Falx cerebri

Septum pellucidum

Corpus callosum

Frontal sinus

Thalamus

Nasal septum

Tentorium cerebelli

Hard palate

Transverse sinus

Oral cavity

Cerebellum

Tongue

Spinal cord

Epiglottis

Esophagus

Larynx

Body of cervical vertebra

Trachea

Intervertebral disc

POSTERIOR

ANTERIOR

INFERIOR

Sagittal section

FIGURE 8.13
Brain and spinal cord

FIGURE 8.14
Brain

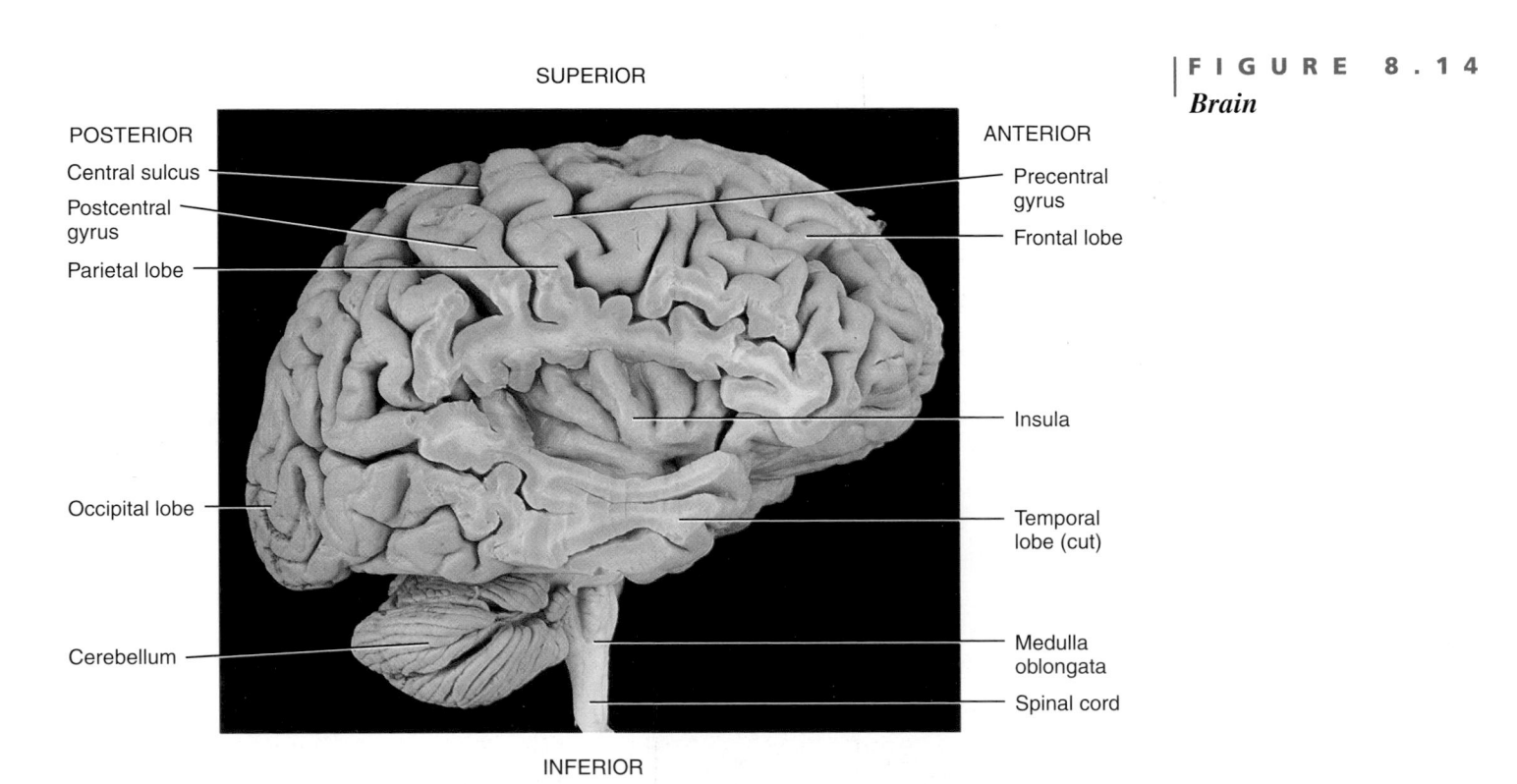

SUPERIOR

POSTERIOR

ANTERIOR

Central sulcus

Precentral gyrus

Postcentral gyrus

Frontal lobe

Parietal lobe

Insula

Occipital lobe

Temporal lobe (cut)

Cerebellum

Medulla oblongata

Spinal cord

INFERIOR

Right lateral view with temporal lobe cut away

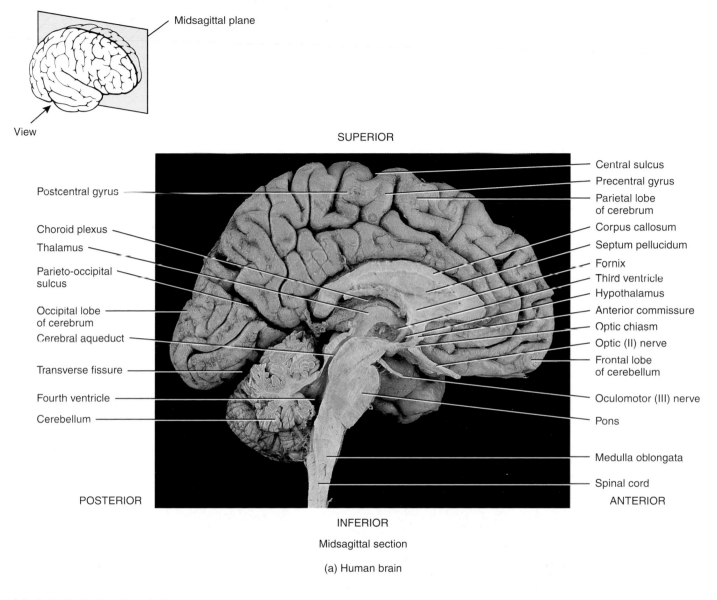

Midsagittal plane

View

SUPERIOR

Central sulcus

Postcentral gyrus

Precentral gyrus

Parietal lobe
of cerebrum

Choroid plexus

Corpus callosum

Thalamus

Septum pellucidum

Parieto-occipital
sulcus

Fornix

Third ventricle

Hypothalamus

Occipital lobe
of cerebrum

Anterior commissure

Cerebral aqueduct

Optic chiasm

Optic (II) nerve

Transverse fissure

Frontal lobe
of cerebellum

Fourth ventricle

Oculomotor (III) nerve

Cerebellum

Pons

Medulla oblongata

Spinal cord

POSTERIOR

ANTERIOR

INFERIOR

Midsagittal section

(a) Human brain

FIGURE 8.15
Brain

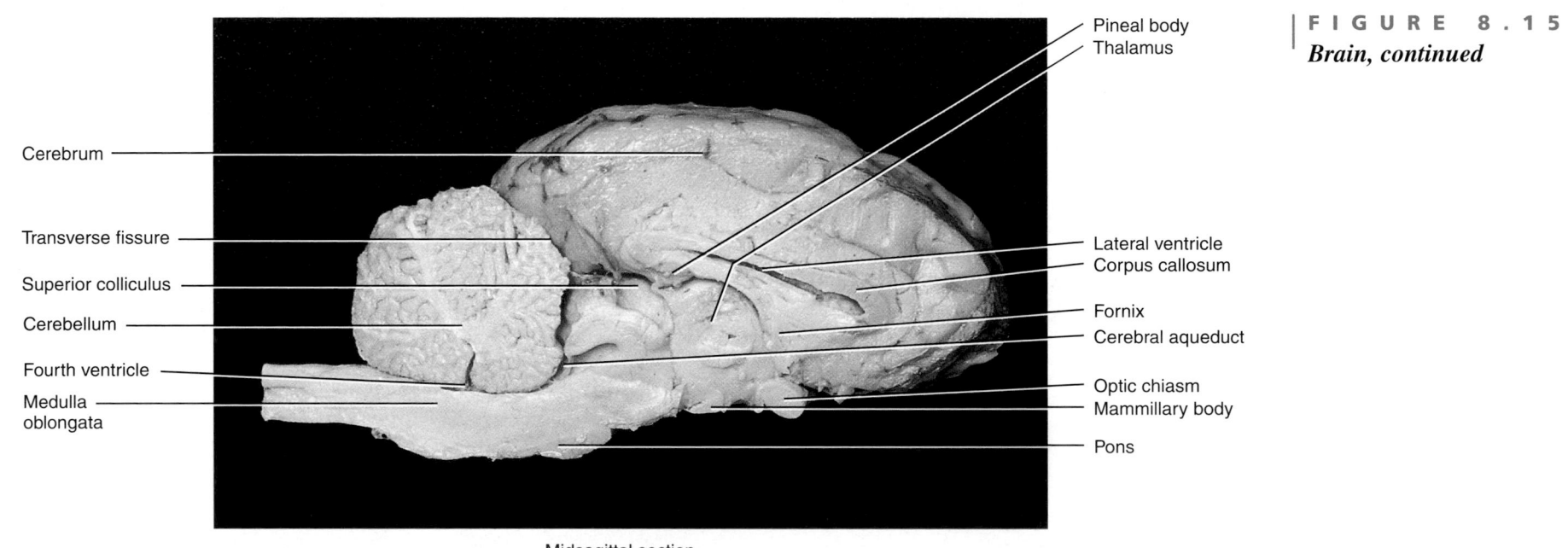

Cerebrum

Transverse fissure

Superior colliculus

Cerebellum

Fourth ventricle

Medulla
oblongata

Pineal body
Thalamus

Lateral ventricle
Corpus callosum

Fornix
Cerebral aqueduct

Optic chiasm
Mammillary body

Pons

Midsagittal section

(b) Sheep brain

FIGURE 8.15
Brain, continued

FIGURE 8.16

White matter tracts of cerebrum revealed by scooping out gray matter of cerebrum

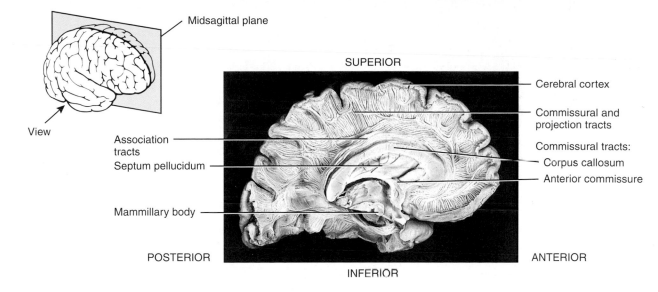

Midsagittal plane

View

SUPERIOR

Cerebral cortex

Commissural and projection tracts

Association tracts

Commissural tracts:

Septum pellucidum

Corpus callosum

Anterior commissure

Mammillary body

POSTERIOR

ANTERIOR

INFERIOR

Midsagittal section

FIGURE 8.17

Basal ganglia revealed by scooping out gray and white matter of cerebrum

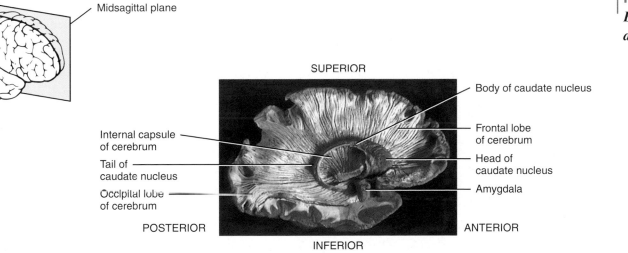

Midsagittal plane

View

SUPERIOR

Body of caudate nucleus

Internal capsule of cerebrum

Frontal lobe of cerebrum

Tail of caudate nucleus

Head of caudate nucleus

Occipital lobe of cerebrum

Amygdala

POSTERIOR

ANTERIOR

INFERIOR

Midsagittal section

FIGURE 8.18
Brain

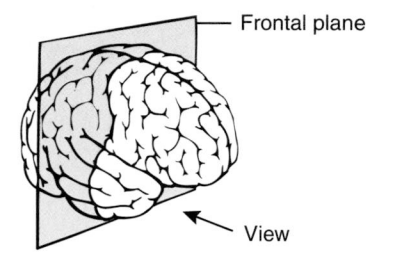

Frontal plane

View

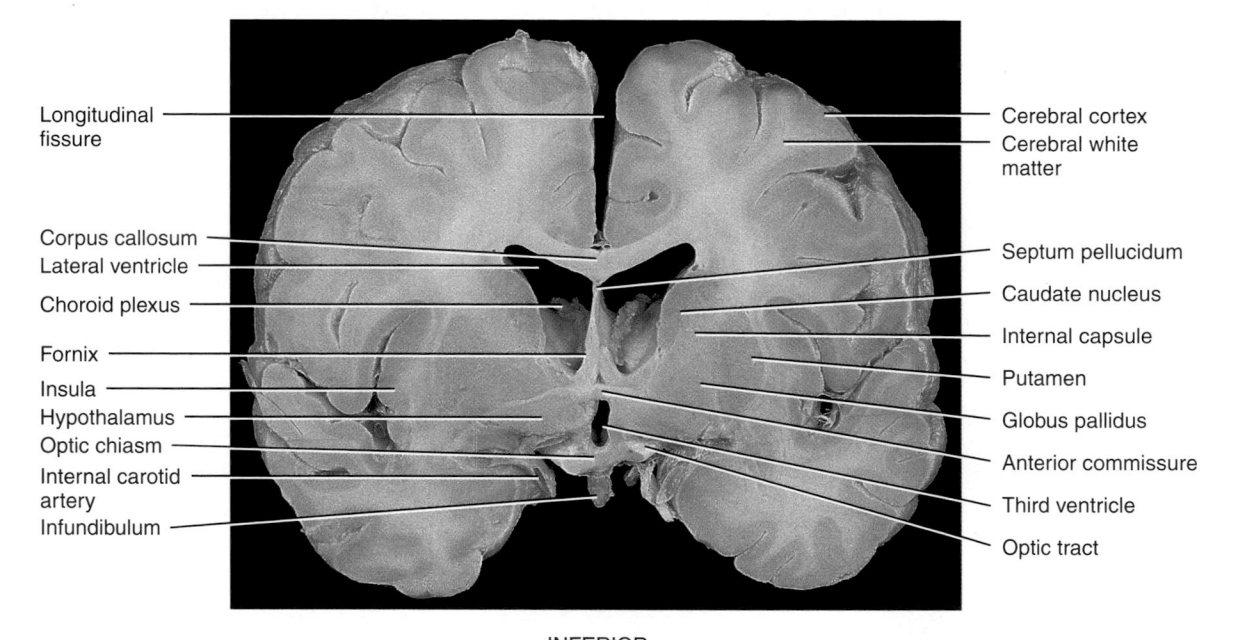

SUPERIOR

Longitudinal fissure

Corpus callosum

Lateral ventricle

Choroid plexus

Fornix

Insula

Hypothalamus

Optic chiasm

Internal carotid artery

Infundibulum

Cerebral cortex

Cerebral white matter

Septum pellucidum

Caudate nucleus

Internal capsule

Putamen

Globus pallidus

Anterior commissure

Third ventricle

Optic tract

INFERIOR

Frontal section

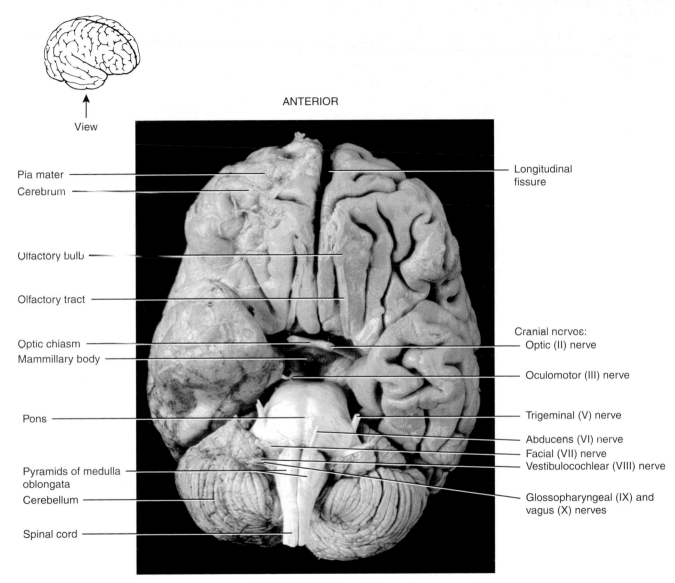

ANTERIOR

View

Pia mater

Cerebrum

Olfactory bulb

Olfactory tract

Optic chiasm

Mammillary body

Pons

Pyramids of medulla oblongata

Cerebellum

Spinal cord

Longitudinal fissure

Cranial nerves:
Optic (II) nerve

Oculomotor (III) nerve

Trigeminal (V) nerve

Abducens (VI) nerve
Facial (VII) nerve
Vestibulocochlear (VIII) nerve

Glossopharyngeal (IX) and vagus (X) nerves

POSTERIOR

(a) Human brain

FIGURE 8.19
Brain

Inferior view

(b) Sheep brain

Labels (b): Olfactory bulb, Olfactory tract, Optic chiasm, Optic tract, Cerebral peduncle, Trigeminal (V) nerve, Facial (VII) nerve, Vestibulocochlear (VIII) nerve, Optic (II) nerve, Pituitary gland, Mammillary body, Pons, Medulla oblongata, Spinal cord

Superior view

(c) Sheep brain

Labels (c): Left cerebral hemisphere, Parietal lobe, Longitudinal fissure, Occipital lobe, Cerebellar hemispheres, Medulla oblongata, Spinal cord, Frontal lobe, Right cerebral hemisphere, Sulci, Gyri, Vermis of cerebellum

FIGURE 8.19
Brain, continued

SUPERIOR

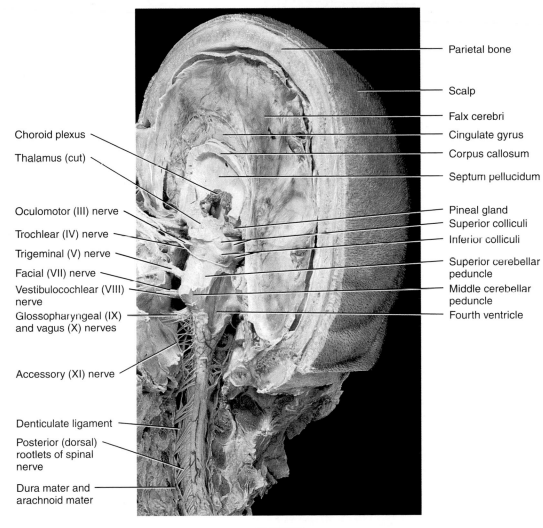

Parietal bone

Scalp

Falx cerebri

Cingulate gyrus

Corpus callosum

Septum pellucidum

Choroid plexus

Thalamus (cut)

Oculomotor (III) nerve

Trochlear (IV) nerve

Trigeminal (V) nerve

Facial (VII) nerve

Vestibulocochlear (VIII) nerve

Glossopharyngeal (IX) and vagus (X) nerves

Accessory (XI) nerve

Denticulate ligament

Posterior (dorsal) rootlets of spinal nerve

Dura mater and arachnoid mater

Pineal gland

Superior colliculi

Inferior colliculi

Superior cerebellar peduncle

Middle cerebellar peduncle

Fourth ventricle

INFERIOR

FIGURE 8.20

Brain (sagittal section) and spinal cord (posterior view)

FIGURE 8.21
Brain

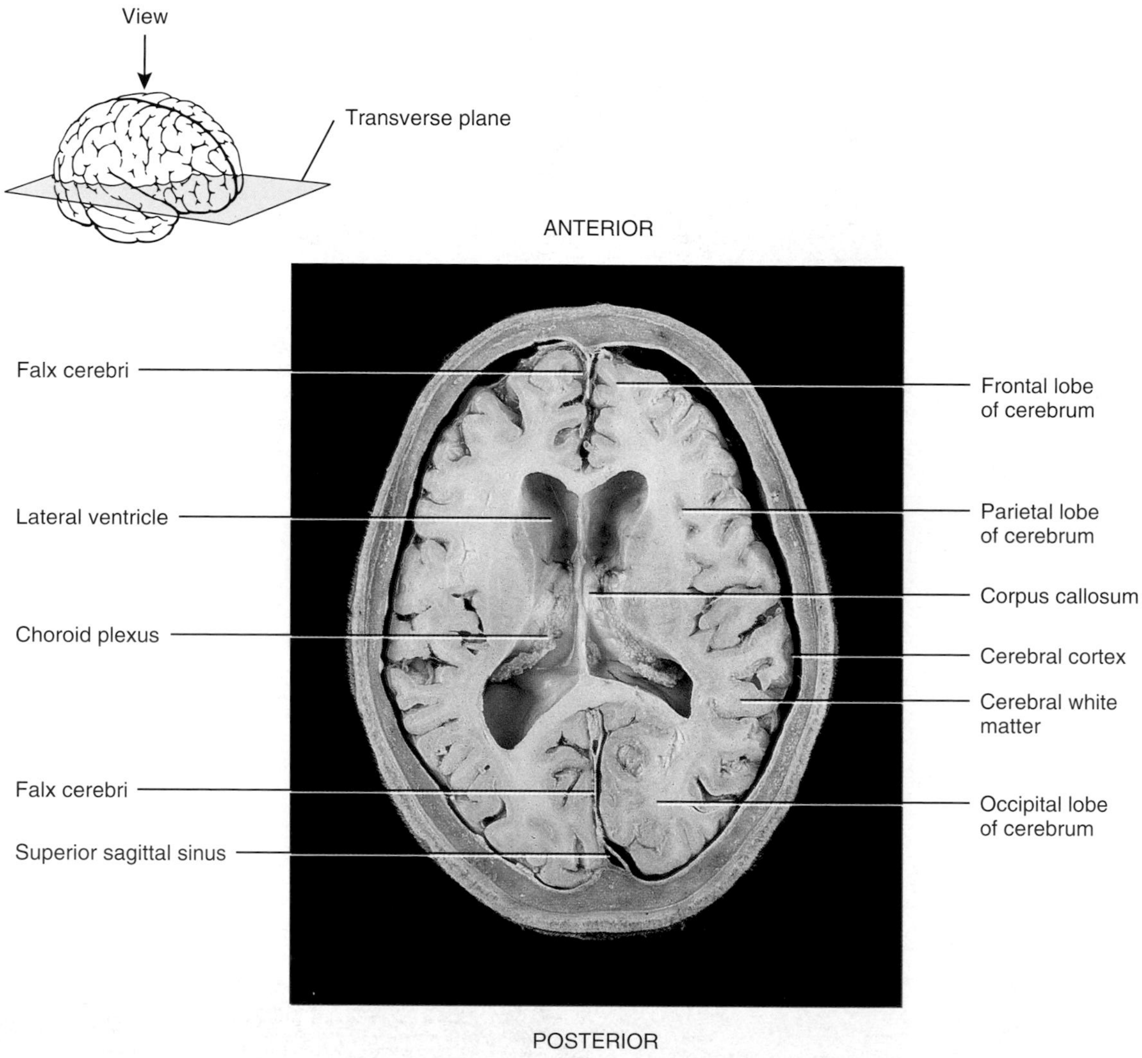
View
Transverse plane
ANTERIOR
Falx cerebri
Lateral ventricle
Choroid plexus
Falx cerebri
Superior sagittal sinus
Frontal lobe of cerebrum
Parietal lobe of cerebrum
Corpus callosum
Cerebral cortex
Cerebral white matter
Occipital lobe of cerebrum
POSTERIOR

Superior view of transverse section

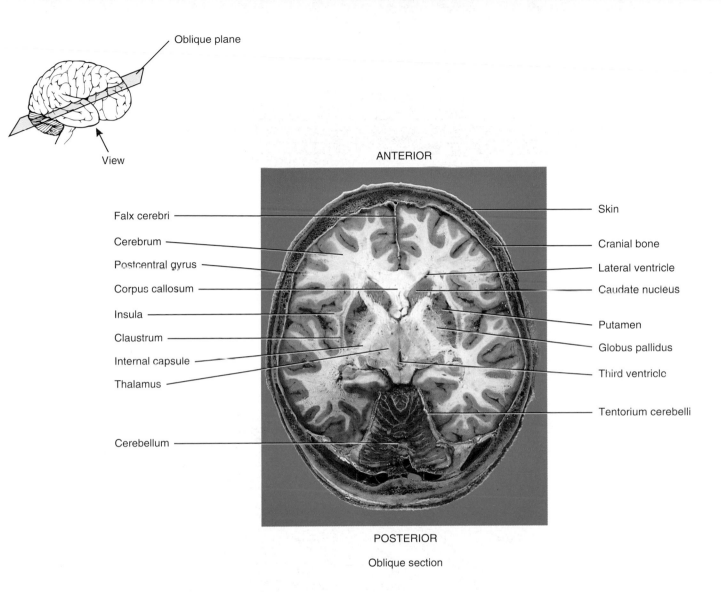

Oblique plane

View

ANTERIOR

Falx cerebri

Cerebrum

Postcentral gyrus

Corpus callosum

Insula

Claustrum

Internal capsule

Thalamus

Cerebellum

Skin

Cranial bone

Lateral ventricle

Caudate nucleus

Putamen

Globus pallidus

Third ventricle

Tentorium cerebelli

POSTERIOR

Oblique section

FIGURE 8.22
Brain

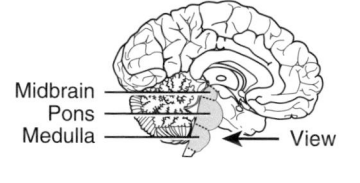

Midbrain
Pons
Medulla ————→ View

SUPERIOR

FIGURE 8.23
Brain stem

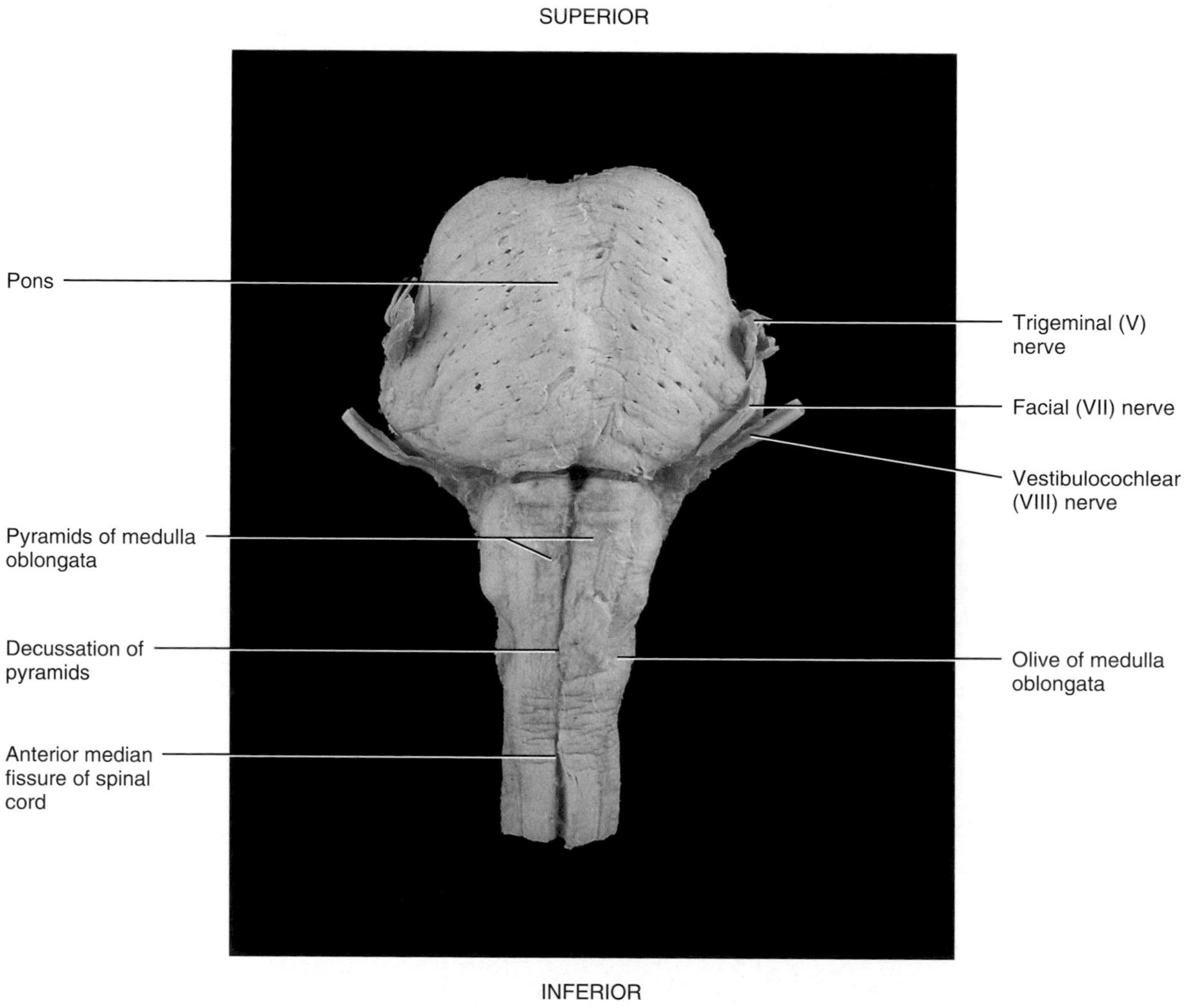

Pons ———————————

Trigeminal (V)
nerve

Facial (VII) nerve

Vestibulocochlear
(VIII) nerve

Pyramids of medulla
oblongata ———————

Decussation of
pyramids ———————

Olive of medulla
oblongata

Anterior median
fissure of spinal
cord ———————

INFERIOR

Anterior view

(a) Human brain

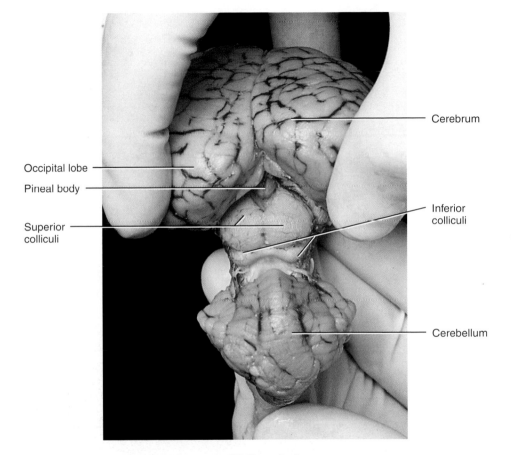

Cerebrum

Occipital lobe

Pineal body

Superior
colliculi

Inferior
colliculi

Cerebellum

(b) Sheep brain

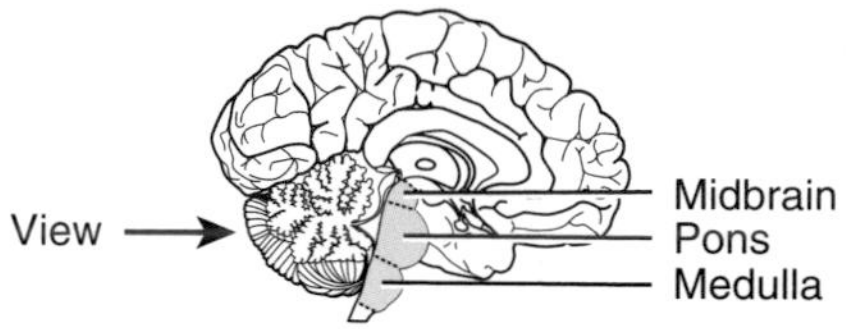

View → Midbrain
Pons
Medulla

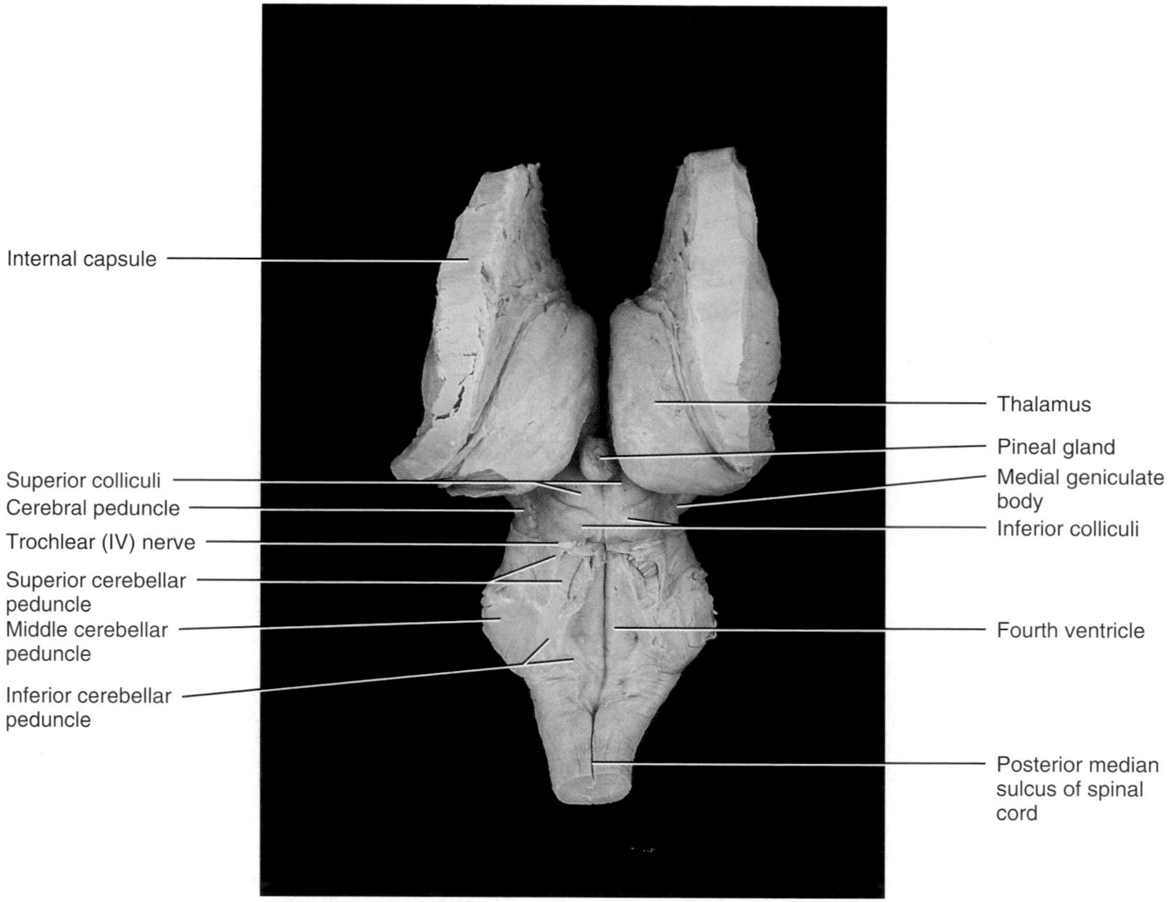

SUPERIOR

Internal capsule

Thalamus

Pineal gland

Superior colliculi
Cerebral peduncle

Medial geniculate
body

Trochlear (IV) nerve

Inferior colliculi

Superior cerebellar
peduncle
Middle cerebellar
peduncle

Fourth ventricle

Inferior cerebellar
peduncle

Posterior median
sulcus of spinal
cord

INFERIOR

Posterior view

FIGURE 8.24
Brain stem

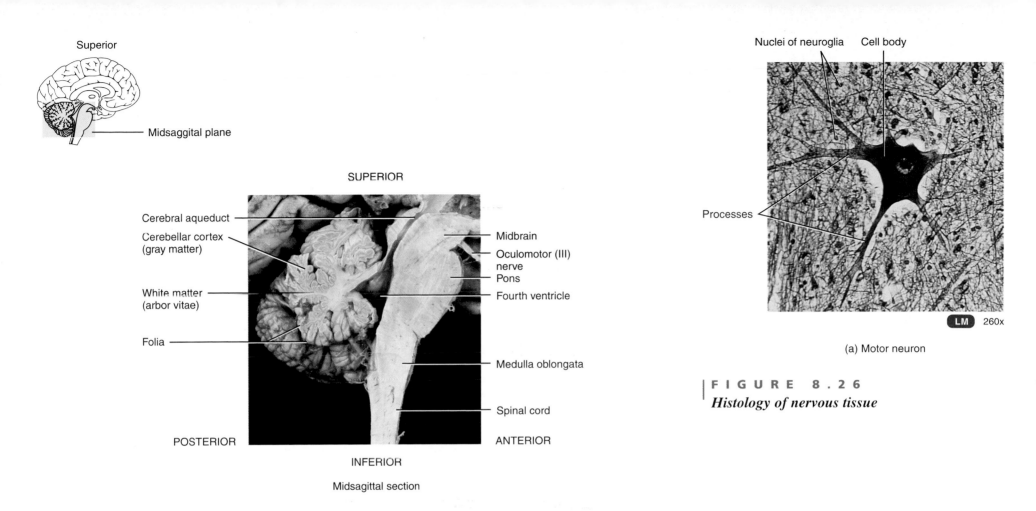

Superior

Midsaggital plane

SUPERIOR

Cerebral aqueduct

Cerebellar cortex
(gray matter)

White matter
(arbor vitae)

Folia

Midbrain

Oculomotor (III)
nerve

Pons

Fourth ventricle

Medulla oblongata

Spinal cord

POSTERIOR

ANTERIOR

INFERIOR

Midsagittal section

FIGURE 8.25
Cerebellum and brain stem

Nuclei of neuroglia   Cell body

Processes

LM   260x

(a) Motor neuron

FIGURE 8.26
Histology of nervous tissue

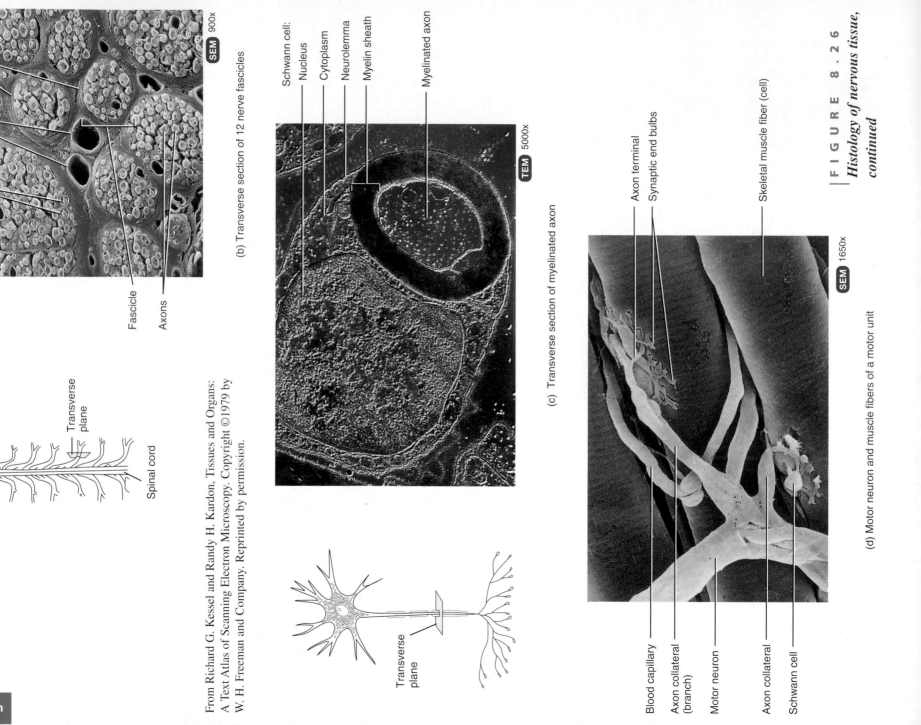

Perineurium

Blood vessels

Endoneurium

Epineurium

Fascicle

Axons

SEM 900x

(b) Transverse section of 12 nerve fascicles

From Richard G. Kessel and Randy H. Kardon, Tissues and Organs: A Text Atlas of Scanning Electron Microscopy. Copyright ©1979 by W. H. Freeman and Company. Reprinted by permission.

Transverse plane

Spinal cord

Schwann cell:
Nucleus
Cytoplasm
Neurolemma
Myelin sheath

Myelinated axon

TEM 5000x

(c) Transverse section of myelinated axon

Transverse plane

Axon terminal
Synaptic end bulbs

Skeletal muscle fiber (cell)

SEM 1650x

Blood capillary

Axon collateral (branch)

Motor neuron

Axon collateral

Schwann cell

(d) Motor neuron and muscle fibers of a motor unit

FIGURE 8.26
Histology of nervous tissue, continued

Bridge

Dorsum nasi

Apex

Philtrum

Root

Ala

External naris

Lips

(a) Anterior view of surface anatomy

SUPERIOR

Brain

Meatuses:
Superior

Middle

Inferior

Soft Palate

Frontal bone

Olfactory bulb

Nasal bone

Conchae:
Superior

Middle

Inferior

Nasal cavity

Hard palate

Oral cavity

INFERIOR

(b) Sagittal section of gross anatomy

Olfactory epithelium

Olfactory gland

Connective tissue

Basal stem cell

Olfactory
receptor cell

Supporting cell

Olfactory hairs

LM 300x

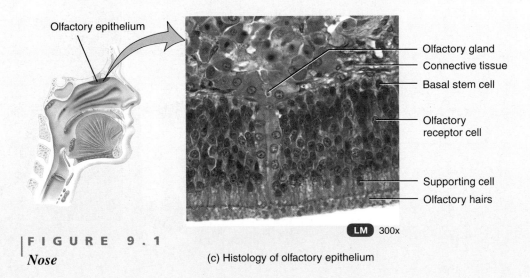

FIGURE 9.1
Nose

(c) Histology of olfactory epithelium

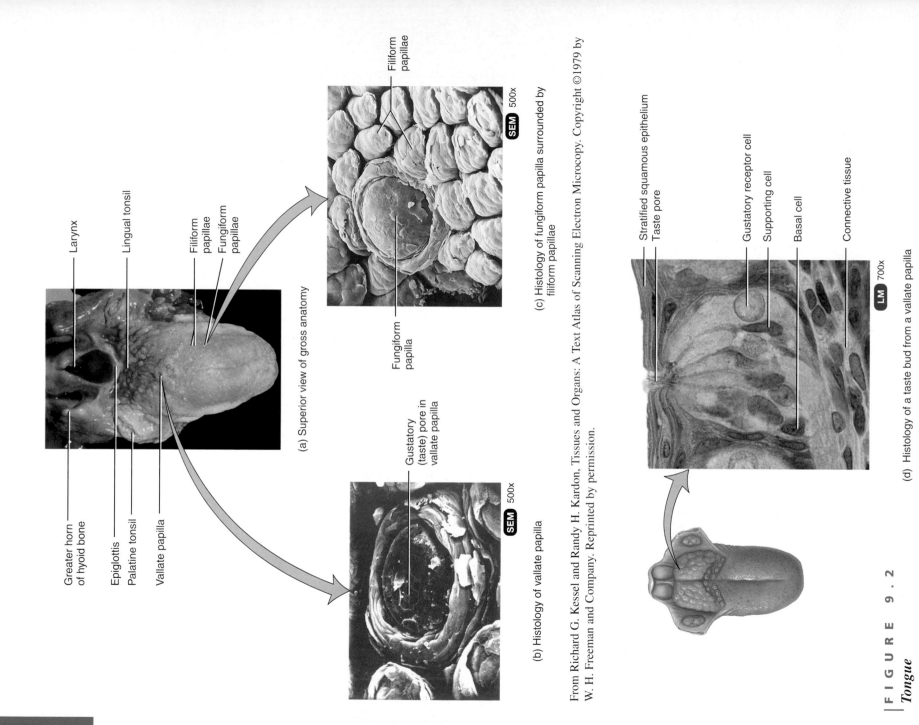

(a) Superior view of gross anatomy

Larynx
Lingual tonsil
Filiform papillae
Fungiform papillae
Greater horn of hyoid bone
Epiglottis
Palatine tonsil
Vallate papilla

(b) Histology of vallate papilla

SEM 500x

Gustatory (taste) pore in vallate papilla

(c) Histology of fungiform papilla surrounded by filiform papillae

SEM 500x

Filiform papillae
Fungiform papilla

From Richard G. Kessel and Randy H. Kardon, Tissues and Organs: A Text Atlas of Scanning Electron Microcopy. Copyright ©1979 by W. H. Freeman and Company. Reprinted by permission.

(d) Histology of a taste bud from a vallate papilla

LM 700x

Stratified squamous epithelium
Taste pore
Gustatory receptor cell
Supporting cell
Basal cell
Connective tissue

FIGURE 9.2
Tongue

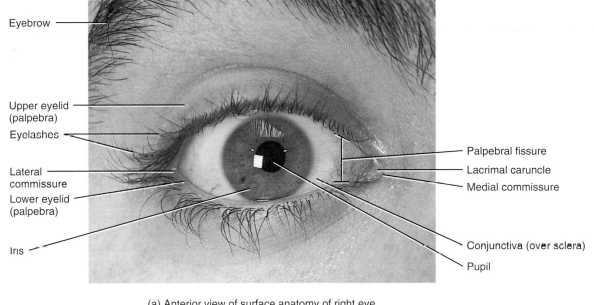

Eyebrow

Upper eyelid
(palpebra)

Eyelashes

Lateral
commissure

Lower eyelid
(palpebra)

Iris

Palpebral fissure

Lacrimal caruncle

Medial commissure

Conjunctiva (over sclera)

Pupil

(a) Anterior view of surface anatomy of right eye

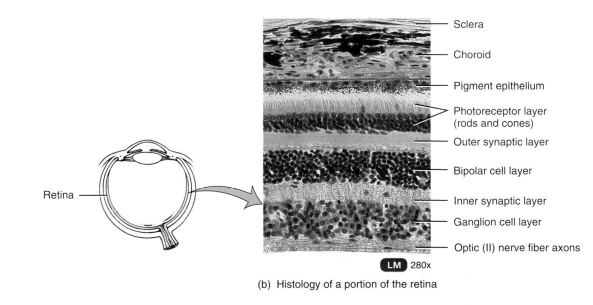

Retina

Sclera

Choroid

Pigment epithelium

Photoreceptor layer
(rods and cones)

Outer synaptic layer

Bipolar cell layer

Inner synaptic layer

Ganglion cell layer

Optic (II) nerve fiber axons

LM 280x

(b) Histology of a portion of the retina

FIGURE 9.3
Eye

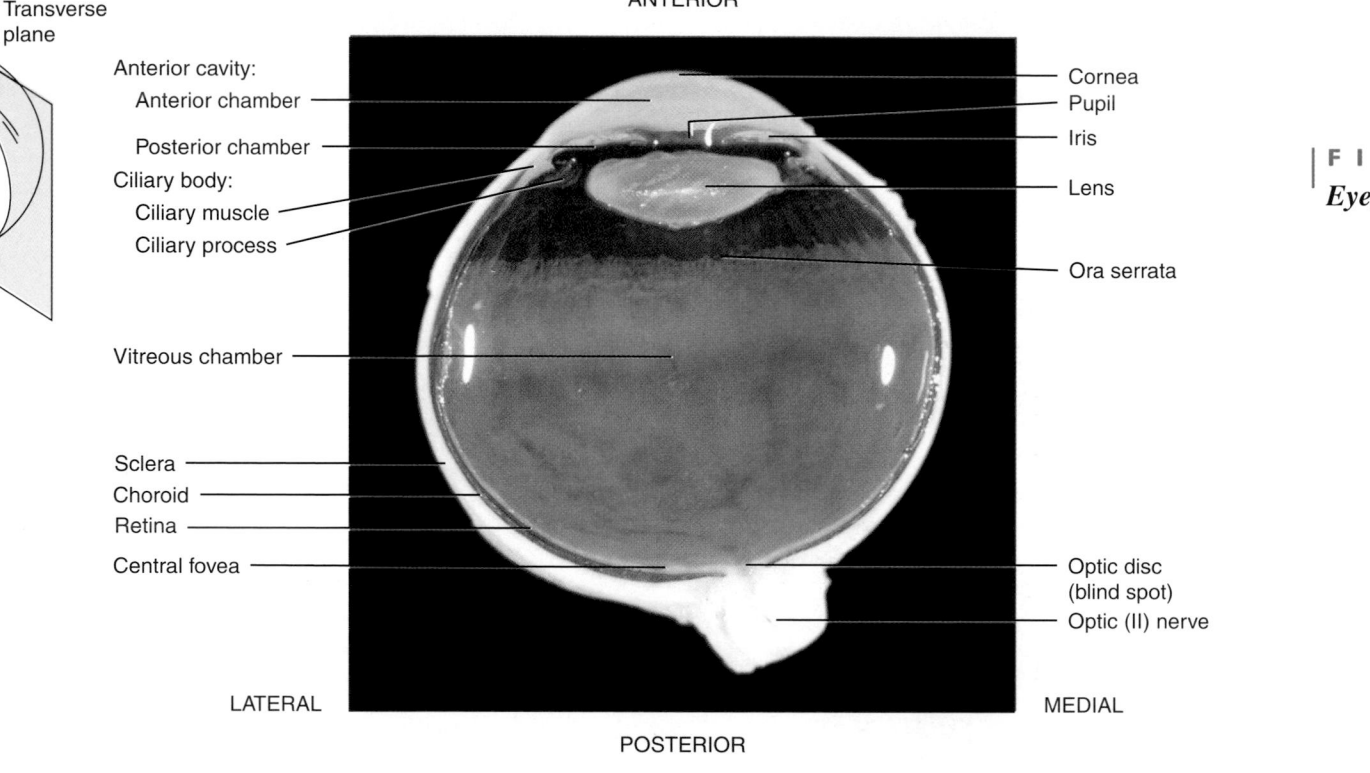

ANTERIOR

Transverse plane

View

Anterior cavity:
  Anterior chamber
  Posterior chamber
Ciliary body:
  Ciliary muscle
  Ciliary process

Vitreous chamber

Sclera
Choroid
Retina
Central fovea

Cornea
Pupil
Iris
Lens

Ora serrata

Optic disc (blind spot)
Optic (II) nerve

LATERAL

MEDIAL

POSTERIOR

(c) Superior view of transverse section of gross anatomy of human left eye

FIGURE 9.3
Eye, continued

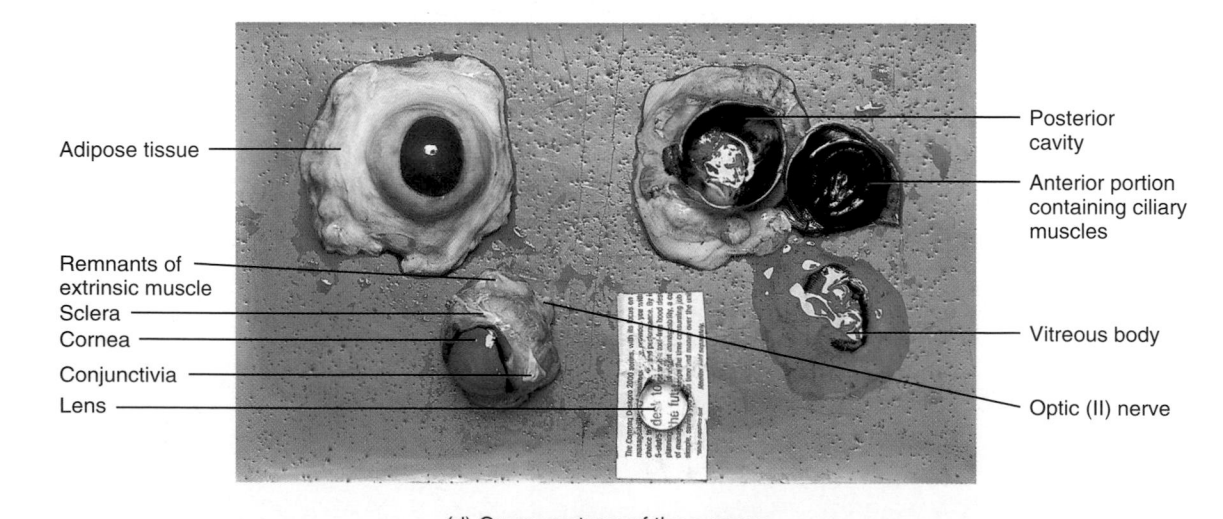

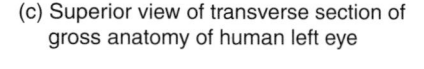

Adipose tissue

Remnants of extrinsic muscle
Sclera
Cornea
Conjunctiva
Lens

Posterior cavity

Anterior portion containing ciliary muscles

Vitreous body

Optic (II) nerve

(d) Gross anatomy of the cow eye

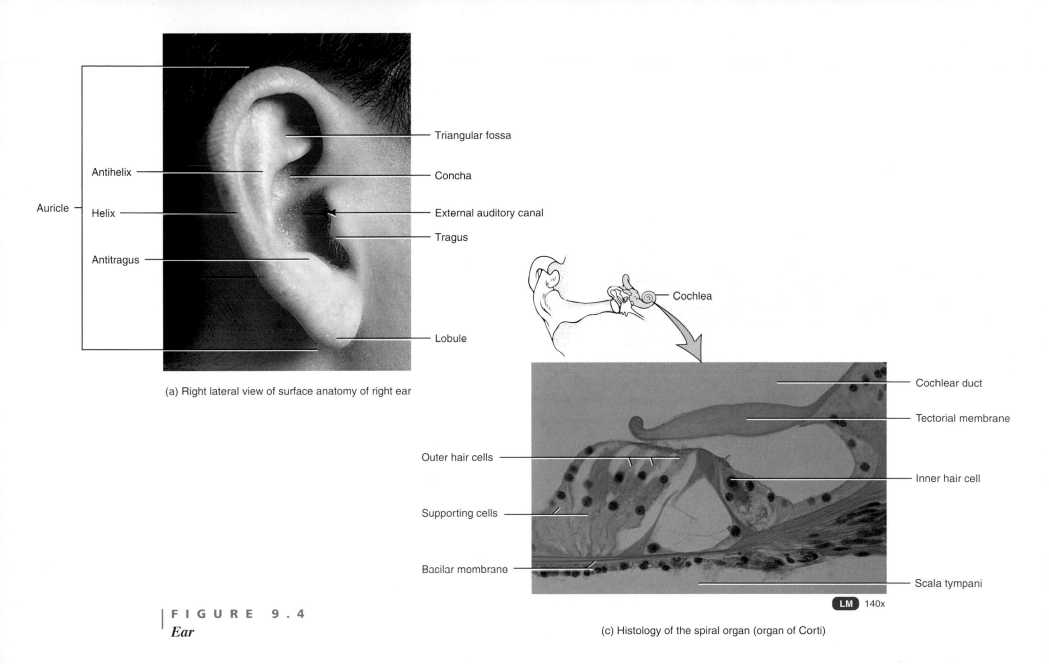

Auricle

Antihelix

Helix

Antitragus

Triangular fossa

Concha

External auditory canal

Tragus

Lobule

(a) Right lateral view of surface anatomy of right ear

Cochlea

Cochlear duct

Tectorial membrane

Outer hair cells

Inner hair cell

Supporting cells

Bacilar mombrane

Scala tympani

LM 140x

(c) Histology of the spiral organ (organ of Corti)

FIGURE 9.4
Ear

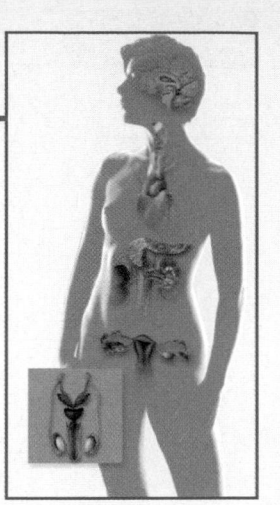

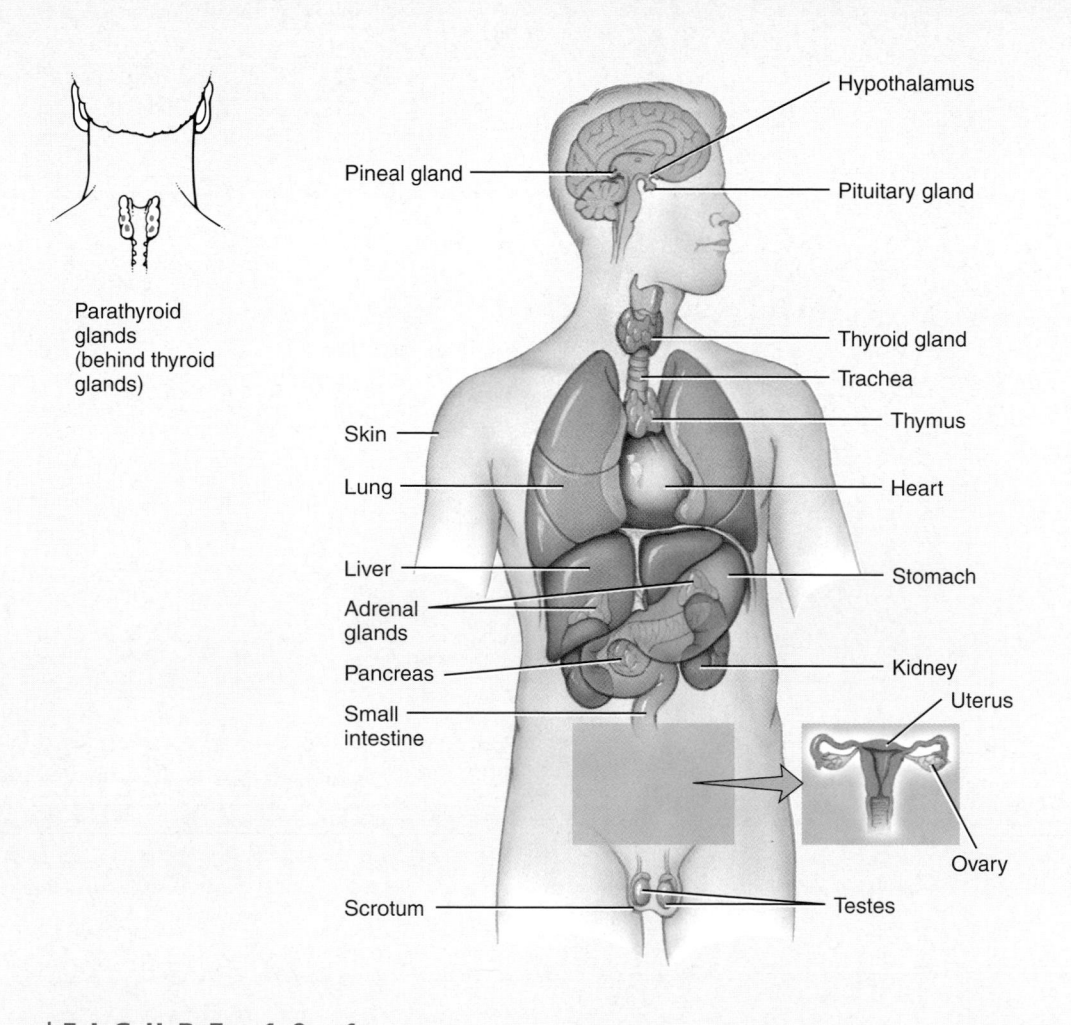

Parathyroid
glands
(behind thyroid
glands)

Pineal gland

Hypothalamus

Pituitary gland

Thyroid gland

Trachea

Thymus

Skin

Lung

Liver

Adrenal
glands

Pancreas

Small
intestine

Scrotum

Heart

Stomach

Kidney

Uterus

Ovary

Testes

FIGURE 10.1

*Endocrine glands, organs containing endocrine tissue,
and associated structures*

FIGURE 10.2
Pituitary gland

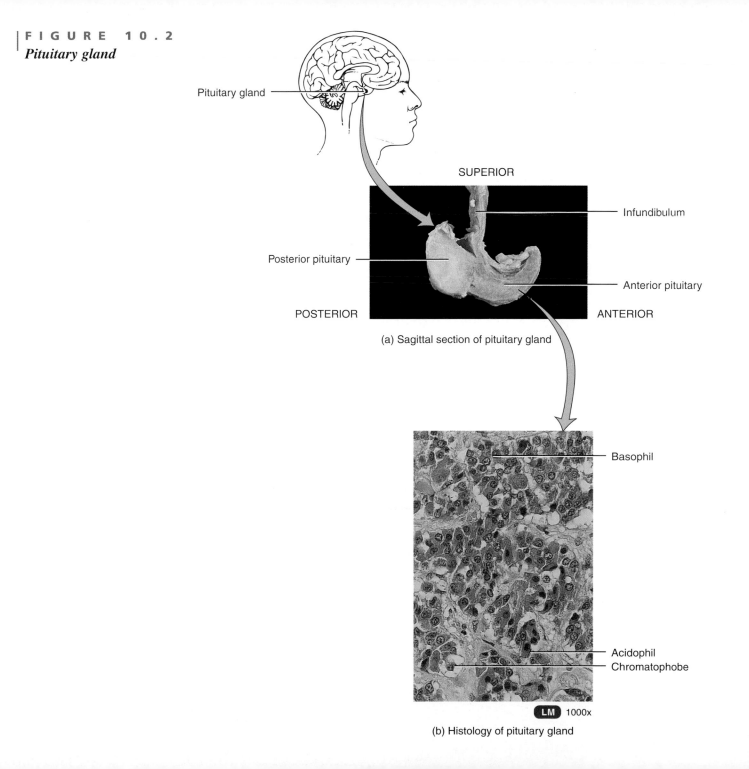

Pituitary gland

SUPERIOR

Infundibulum

Posterior pituitary

Anterior pituitary

POSTERIOR

ANTERIOR

(a) Sagittal section of pituitary gland

Basophil

Acidophil
Chromatophobe

LM 1000x

(b) Histology of pituitary gland

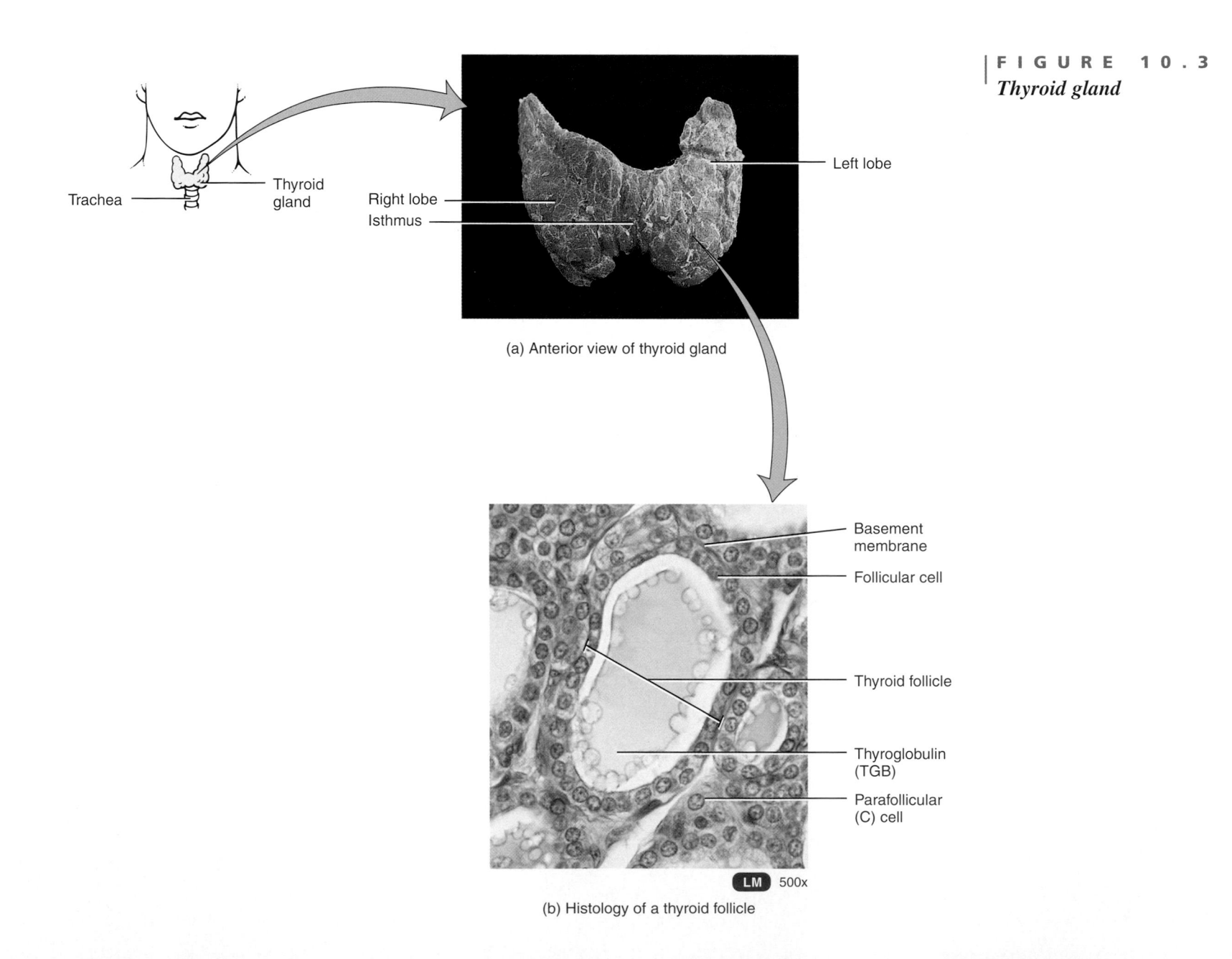

FIGURE 10.3
Thyroid gland

Trachea

Thyroid gland

Right lobe

Isthmus

Left lobe

(a) Anterior view of thyroid gland

Basement membrane

Follicular cell

Thyroid follicle

Thyroglobulin (TGB)

Parafollicular (C) cell

LM 500x

(b) Histology of a thyroid follicle

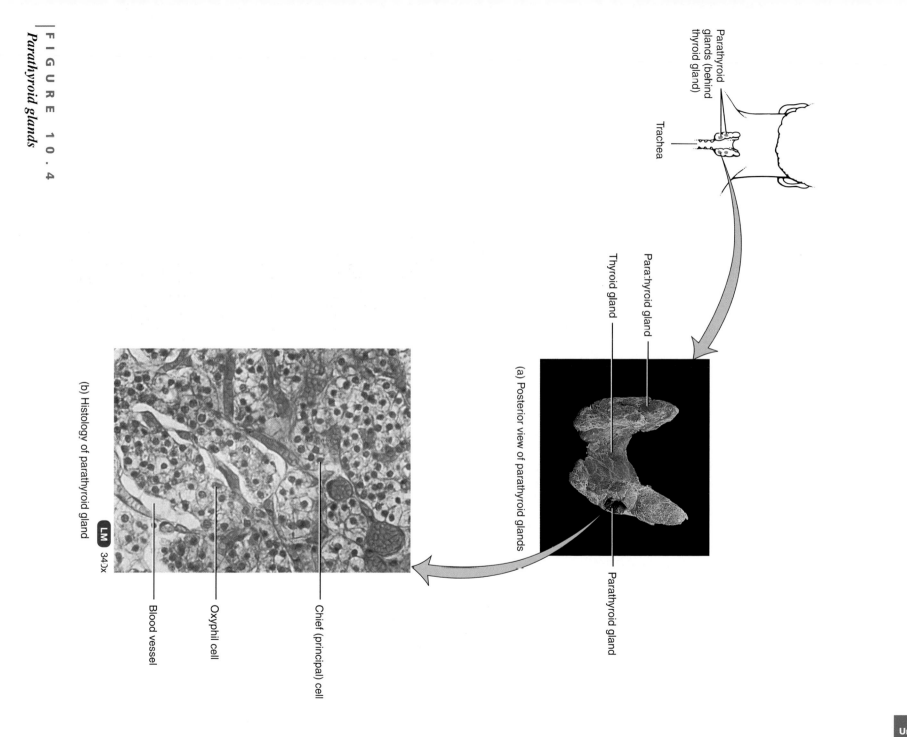

Parathyroid
glands (behind
thyroid gland)

Trachea

Thyroid gland

Parathyroid gland

(a) Posterior view of parathyroid glands

Parathyroid gland

(b) Histology of parathyroid gland

LM 340x

Blood vessel

Oxyphil cell

Chief (principal) cell

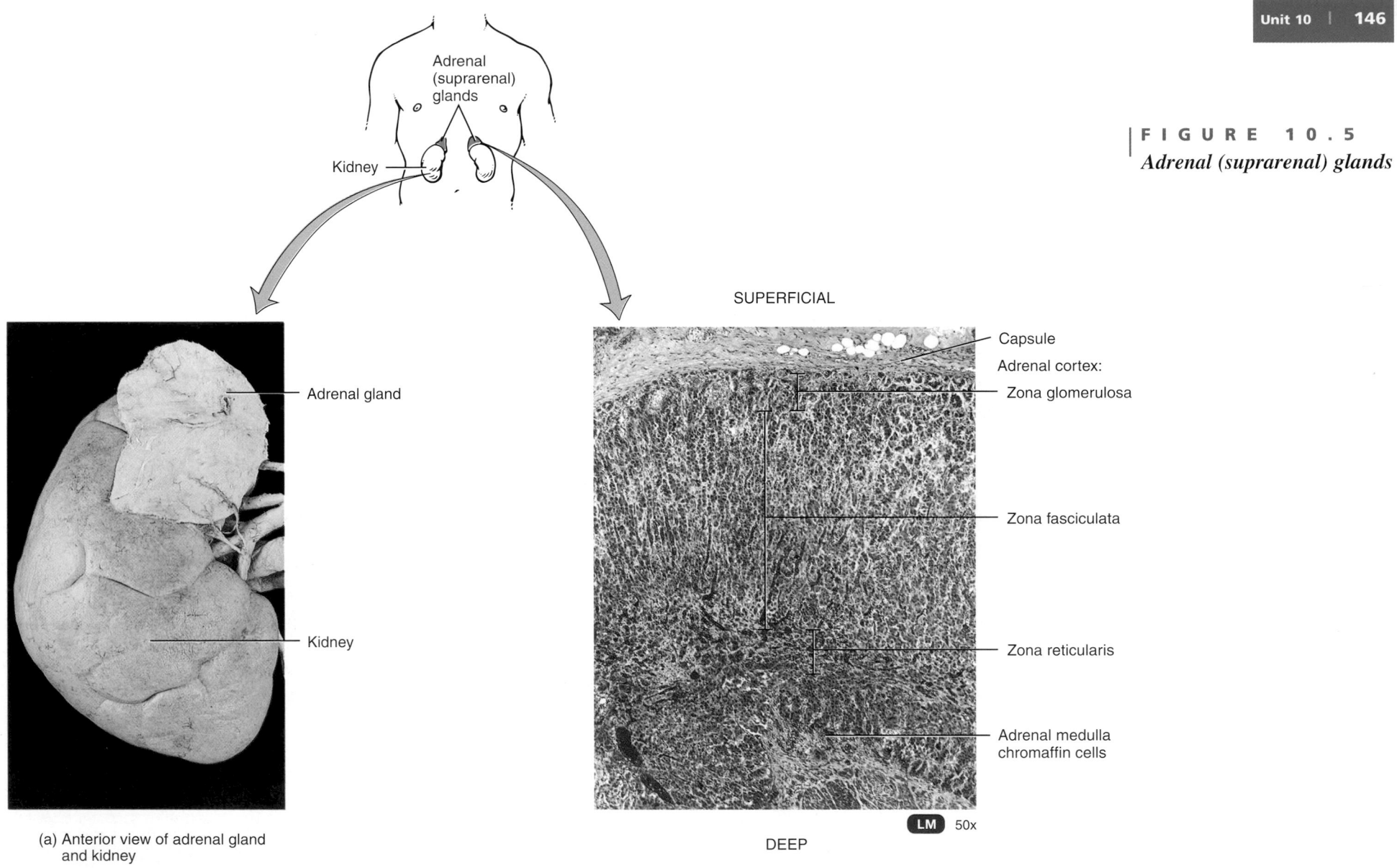

Adrenal (suprarenal) glands

Kidney

FIGURE 10.5
Adrenal (suprarenal) glands

Adrenal gland

Kidney

(a) Anterior view of adrenal gland
and kidney

SUPERFICIAL

Capsule
Adrenal cortex:
Zona glomerulosa

Zona fasciculata

Zona reticularis

Adrenal medulla
chromaffin cells

LM 50x

DEEP

(b) Histology of adrenal gland

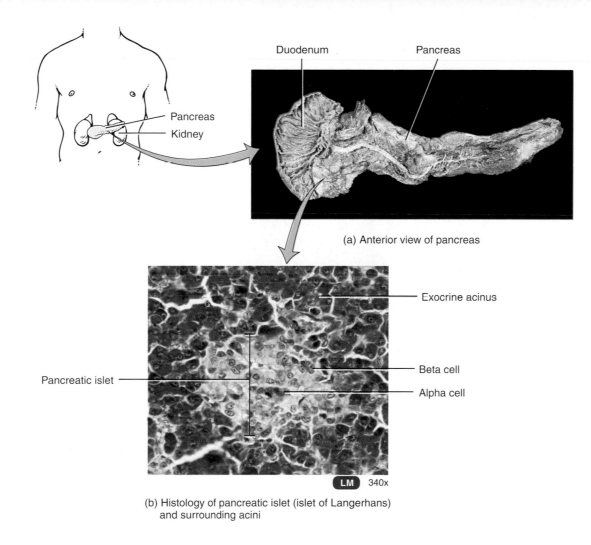

Duodenum   Pancreas

(a) Anterior view of pancreas

Exocrine acinus

Beta cell

Alpha cell

Pancreatic islet

LM   340x

(b) Histology of pancreatic islet (islet of Langerhans)
and surrounding acini

Pancreas
Kidney

FIGURE 10.6
Pancreas

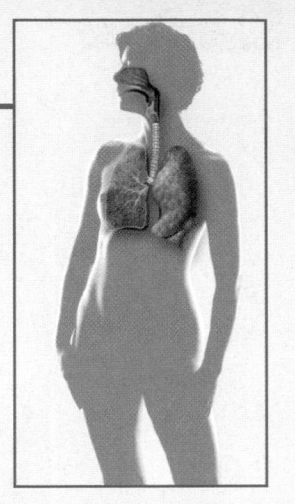

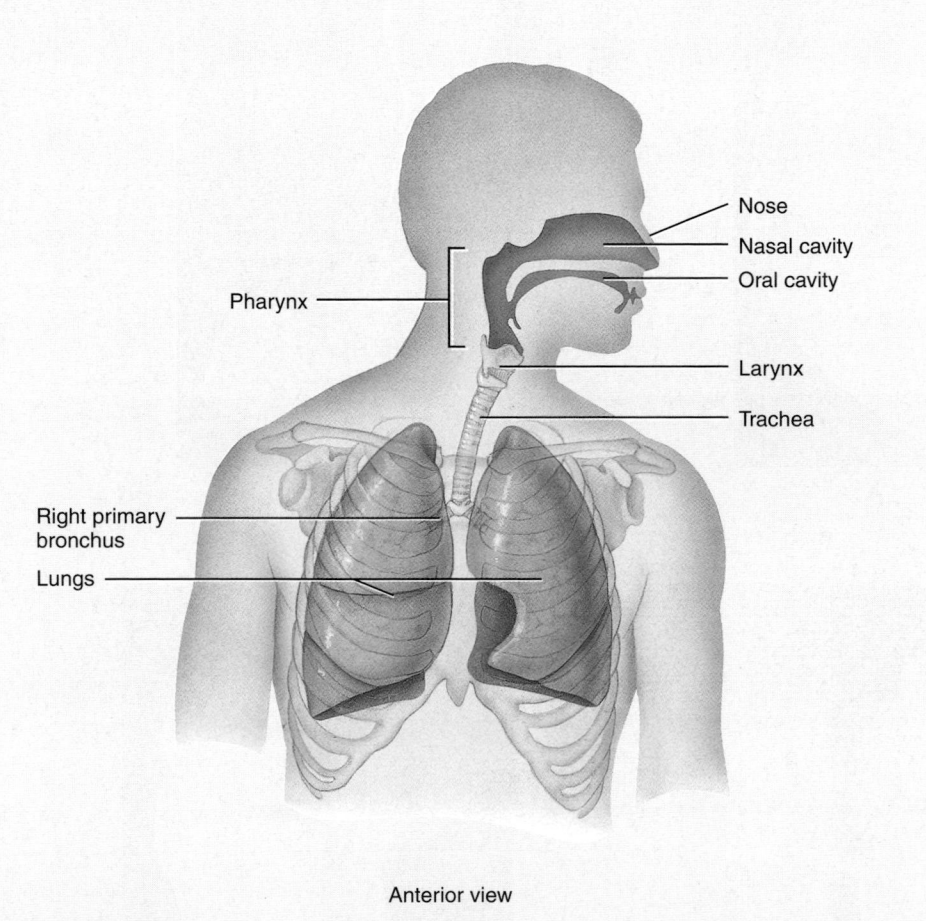

Pharynx

Nose

Nasal cavity

Oral cavity

Larynx

Trachea

Right primary
bronchus

Lungs

Anterior view

FIGURE 11.1
Respiratory System

Sagittal plane

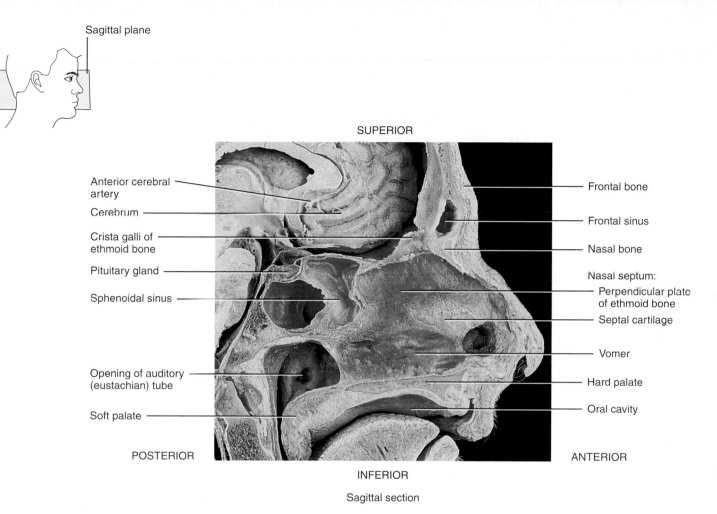

SUPERIOR

Anterior cerebral artery

Cerebrum

Crista galli of ethmoid bone

Pituitary gland

Sphenoidal sinus

Opening of auditory (eustachian) tube

Soft palate

Frontal bone

Frontal sinus

Nasal bone

Nasal septum:
Perpendicular plate of ethmoid bone

Septal cartilage

Vomer

Hard palate

Oral cavity

POSTERIOR

ANTERIOR

INFERIOR

Sagittal section

FIGURE 11.2
Nasal septum

FIGURE 11.3
Nasal cavity

Sagittal plane

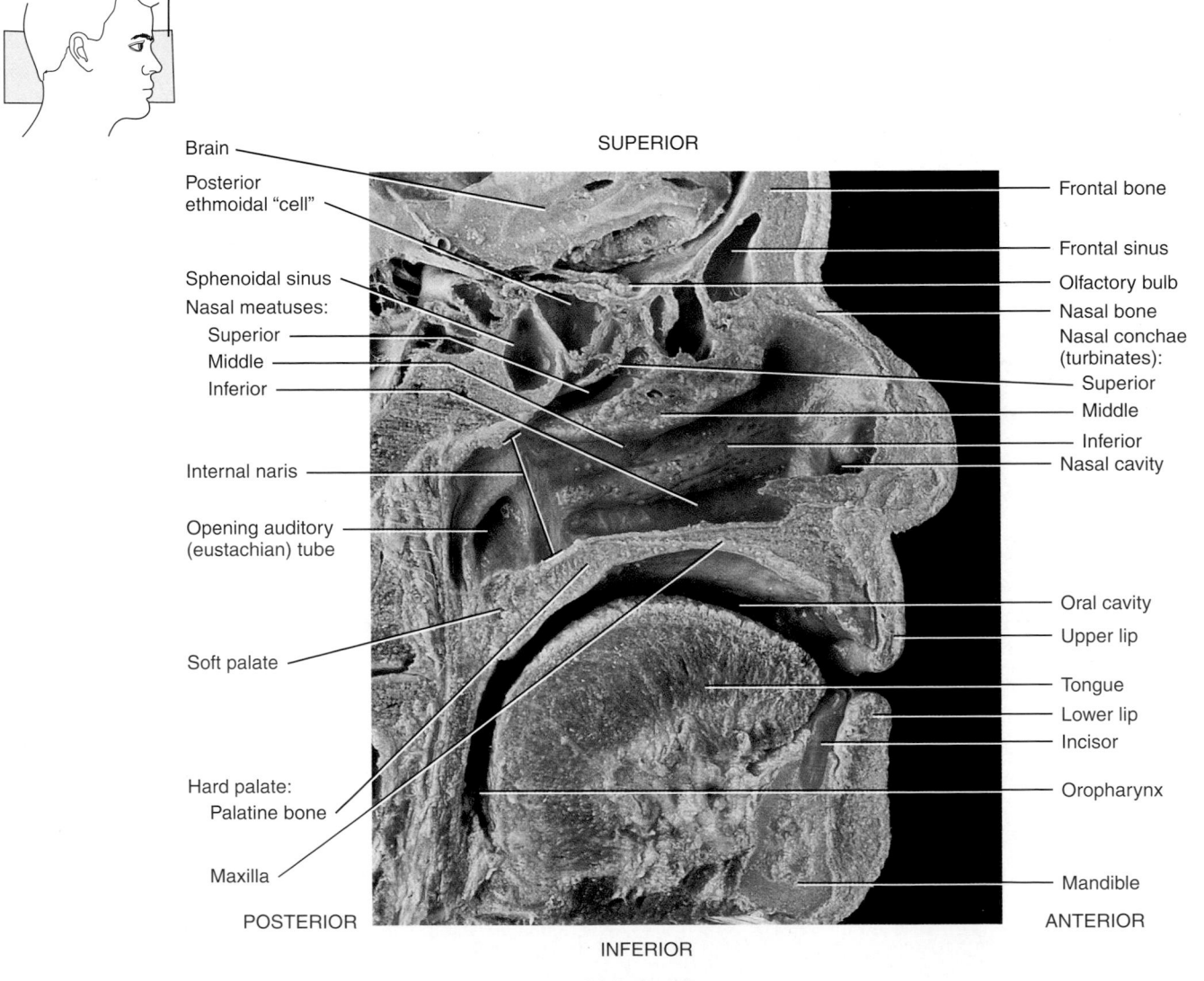

SUPERIOR

Brain

Posterior
ethmoidal "cell"

Frontal bone

Frontal sinus

Sphenoidal sinus

Nasal meatuses:

   Superior

   Middle

   Inferior

Olfactory bulb

Nasal bone

Nasal conchae
(turbinates):

   Superior

   Middle

   Inferior

Internal naris

Nasal cavity

Opening auditory
(eustachian) tube

Oral cavity

Upper lip

Soft palate

Tongue

Lower lip

Incisor

Hard palate:
   Palatine bone

Oropharynx

Maxilla

Mandible

POSTERIOR

ANTERIOR

INFERIOR

Sagittal section

Sagittal plane

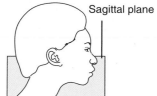

Regions of the pharynx

Nasopharynx

Oropharynx

Laryngopharynx

Regions of the pharynx

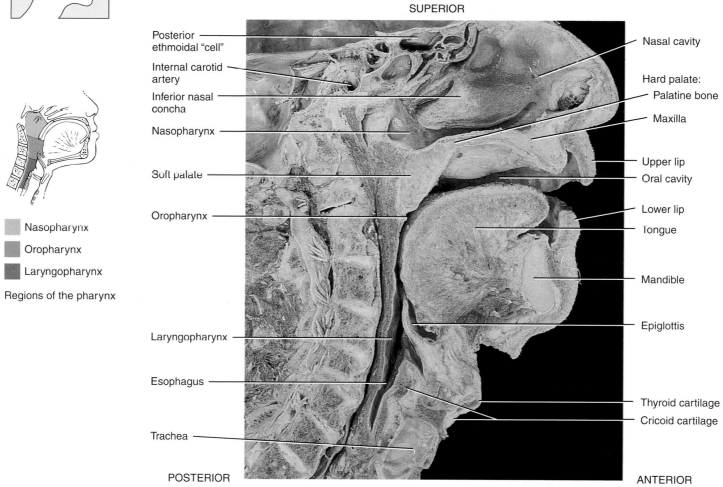

SUPERIOR

Posterior ethmoidal "cell"

Internal carotid artery

Inferior nasal concha

Nasopharynx

Soft palate

Oropharynx

Laryngopharynx

Esophagus

Trachea

POSTERIOR

Nasal cavity

Hard palate:
Palatine bone

Maxilla

Upper lip

Oral cavity

Lower lip

Tongue

Mandible

Epiglottis

Thyroid cartilage

Cricoid cartilage

ANTERIOR

INFERIOR

Sagittal section

F I G U R E   1 1 . 4
Subdivisions of pharynx

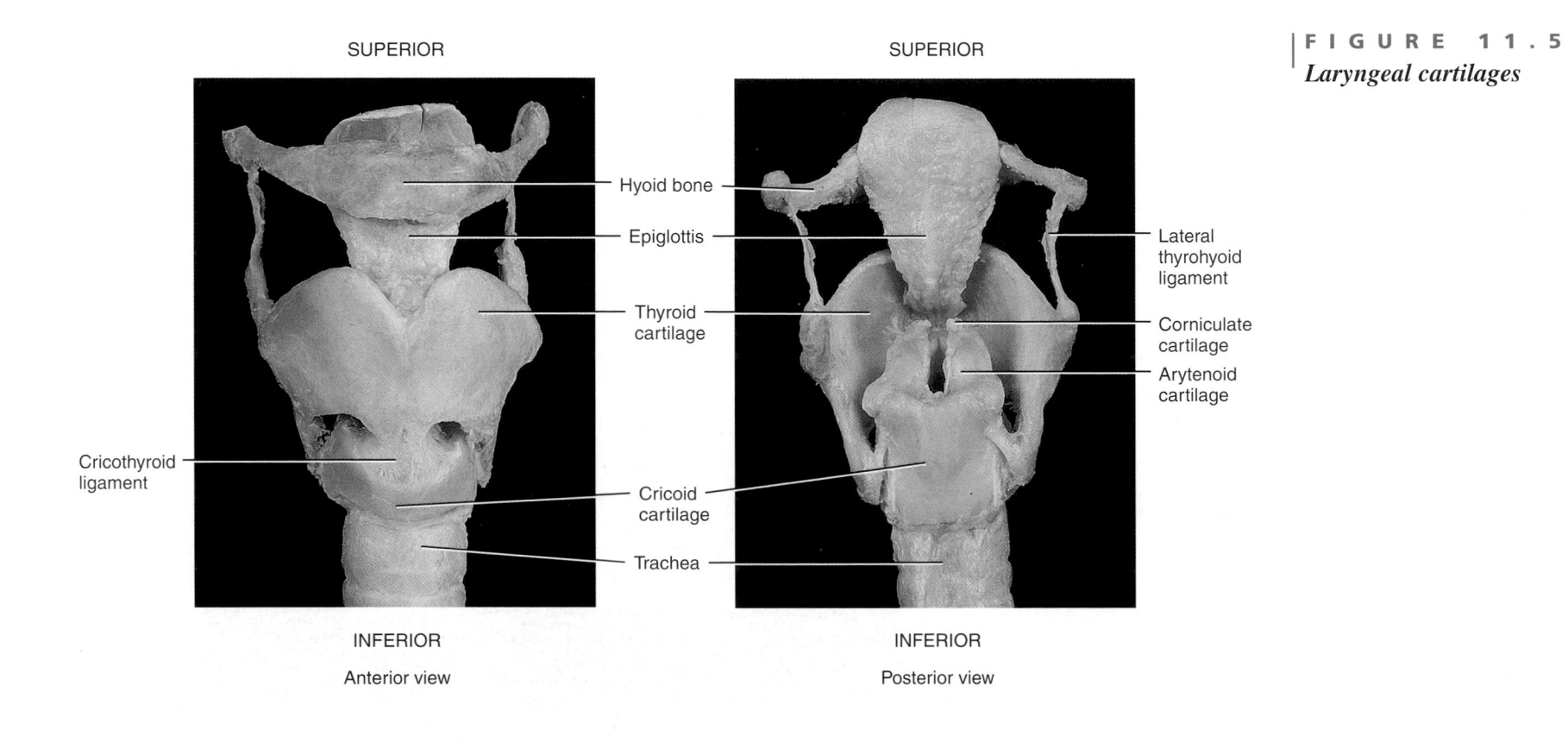

SUPERIOR

SUPERIOR

Hyoid bone

Epiglottis

Lateral
thyrohyoid
ligament

Thyroid
cartilage

Corniculate
cartilage

Arytenoid
cartilage

Cricothyroid
ligament

Cricoid
cartilage

Trachea

INFERIOR

INFERIOR

Anterior view

Posterior view

FIGURE 11.5
Laryngeal cartilages

SUPERIOR

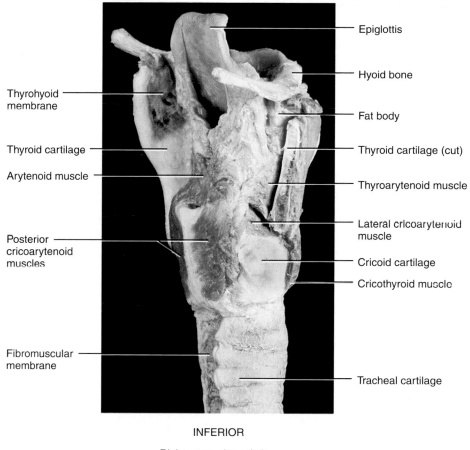

Epiglottis

Hyoid bone

Thyrohyoid
membrane

Fat body

Thyroid cartilage

Thyroid cartilage (cut)

Arytenoid muscle

Thyroarytenoid muscle

Lateral cricoarytenoid
muscle

Posterior
cricoarytenoid
muscles

Cricoid cartilage

Cricothyroid muscle

Fibromuscular
membrane

Tracheal cartilage

INFERIOR

Right posterolateral view

FIGURE 11.6
Larynx

FIGURE 11.7
Larynx

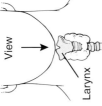

View →

Larynx

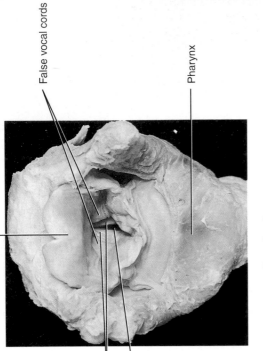

Epiglottis

False vocal cords

Pharynx

Glottis:
True vocal cords
Rima glottidis

Superior view

SUPERIOR

Hyoid bone

Thyroid cartilage of larynx

Cricoid cartilage of larynx

Trachea

Left primary bronchus

Left secondary bronchus

Left tertiary bronchus

Left bronchiole

Left terminal bronchiole

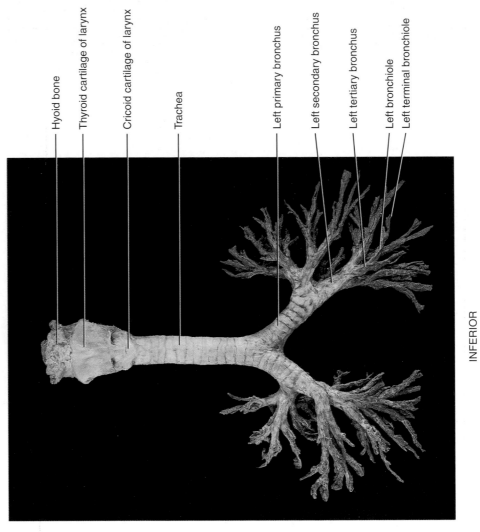

INFERIOR

Anterior view

BRANCHING OF
BRONCHIAL TREE

Trachea
↓
Primary bronchi
↓
Secondary bronchi
↓
Tertiary bronchi
↓
Bronchioles
↓
Terminal bronchioles

FIGURE 11.8
Bronchial tree

F I G U R E 1 1 . 9

Histology of the trachea

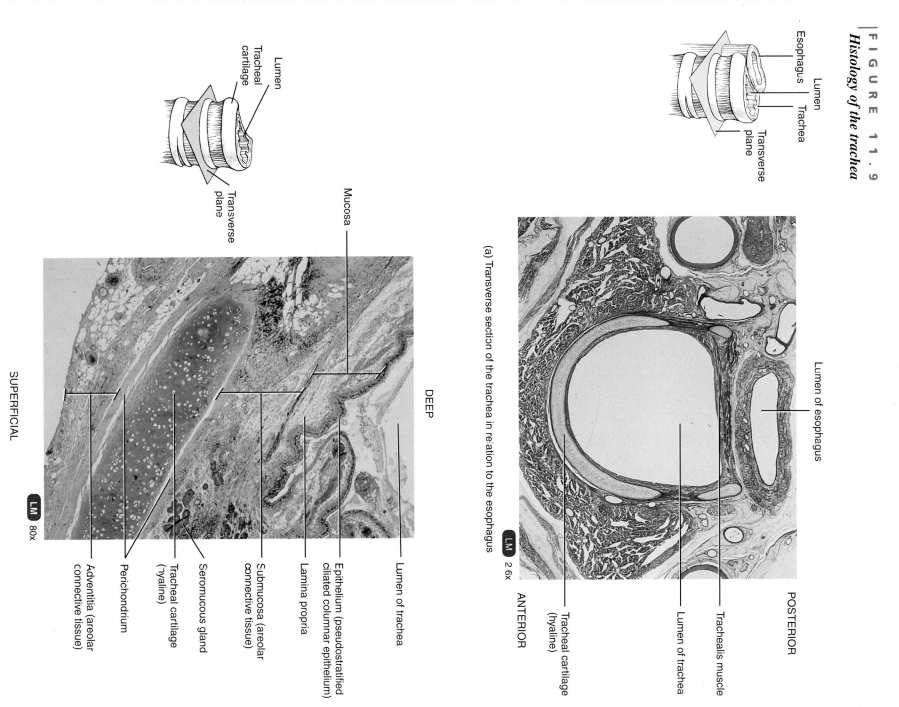

Esophagus

Lumen

Trachea

Transverse plane

Lumen of esophagus

Tracheal cartilage (hyaline)

Lumen of trachea

Trachealis muscle

POSTERIOR

ANTERIOR

LM 2.6x

(a) Transverse section of the trachea in relation to the esophagus

Lumen

Tracheal cartilage

Transverse plane

Mucosa

SUPERFICIAL

DEEP

Adventitia (areolar connective tissue)

Perichondrium

Tracheal cartilage (hyaline)

Seromucous gland

Submucosa (areolar connective tissue)

Lamina propria

Epithelium (pseudostratified ciliated columnar epithelium)

Lumen of trachea

LM 80x

(b) Transverse section of part of the tracheal wall

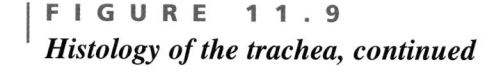

FIGURE 11.9
Histology of the trachea, continued

Mucus in goblet cell   Cilia   Lumen

Nucleus of ciliated
columnar cell

Nucleus of goblet cell

Nucleus of basal cell

Basement membrane

Lamina propria
(areolar connective tissue)

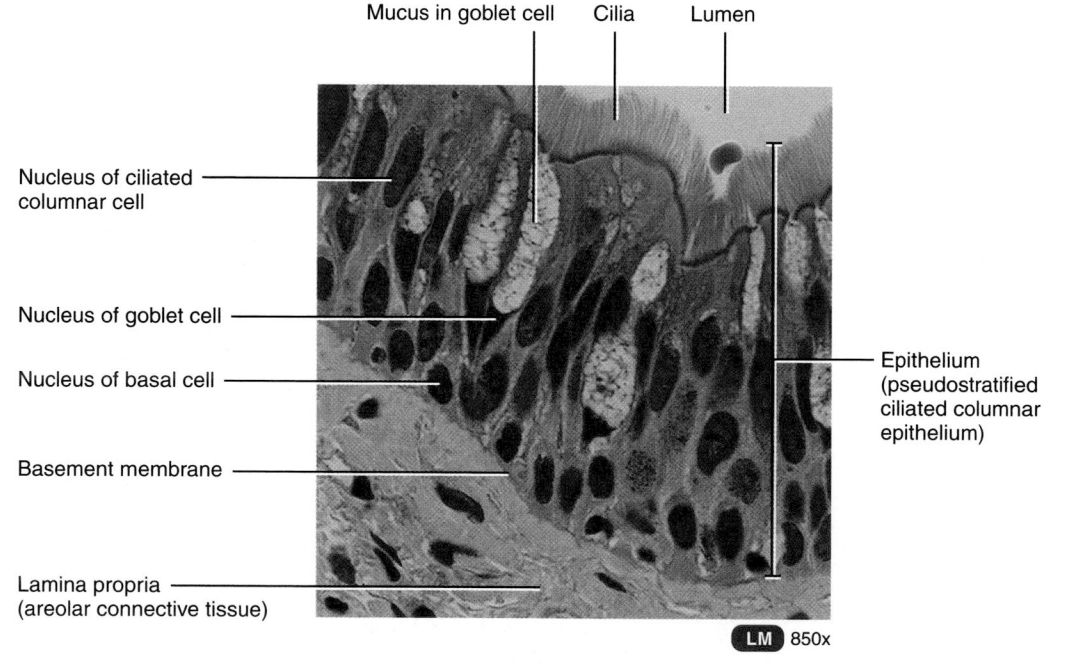

Epithelium
(pseudostratified
ciliated columnar
epithelium)

LM 850x

(c) Transverse section of tracheal epithelium

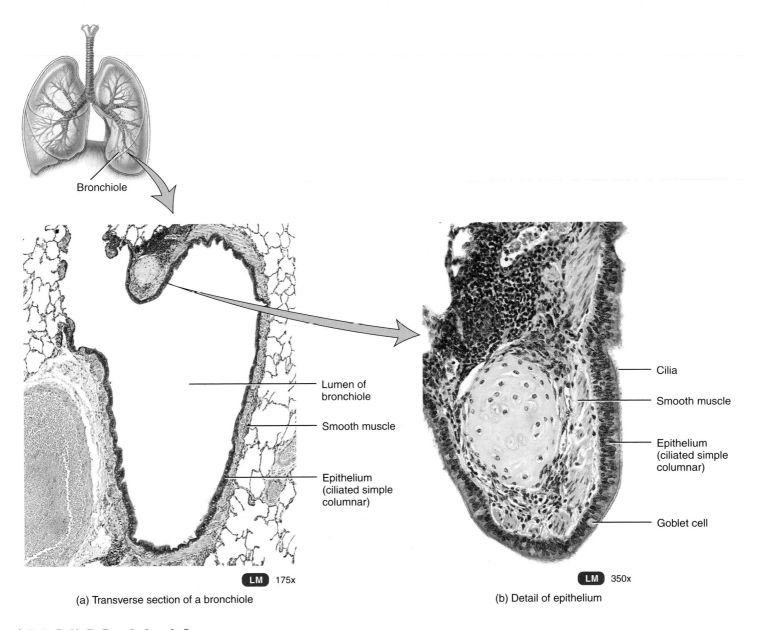

Bronchiole

Lumen of
bronchiole

Smooth muscle

Epithelium
(ciliated simple
columnar)

LM 175x

(a) Transverse section of a bronchiole

Cilia

Smooth muscle

Epithelium
(ciliated simple
columnar)

Goblet cell

LM 350x

(b) Detail of epithelium

FIGURE 11.10
Histology of a bronchiole

FIGURE 11.11
Lungs

SUPERIOR

Trachea

Right internal
jugular vein

Right common
carotid artery

Right axillary
vein

Right
brachiocephalic
vein

Rib

Right lung

Diaphragm

Right lobe
of liver

Thyroid gland

Left internal
jugular vein

Left common
carotid artery

Left
brachiocephalic
vein

Left lung

Heart

Cut edge of
pericardium

Diaphragm

Left lobe
of liver

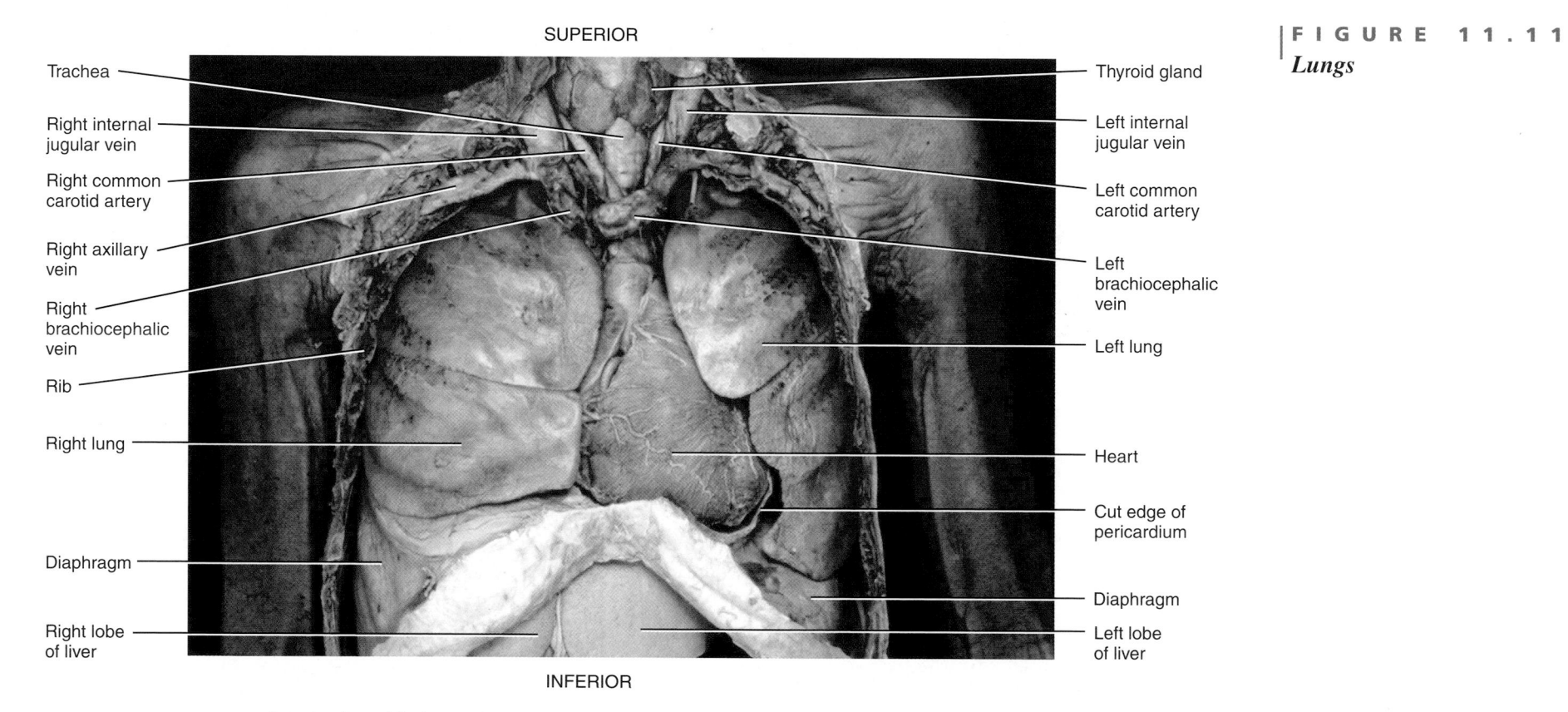

INFERIOR

Anterior view of the lungs after removal of the anterolateral thoracic wall and parietal pleura

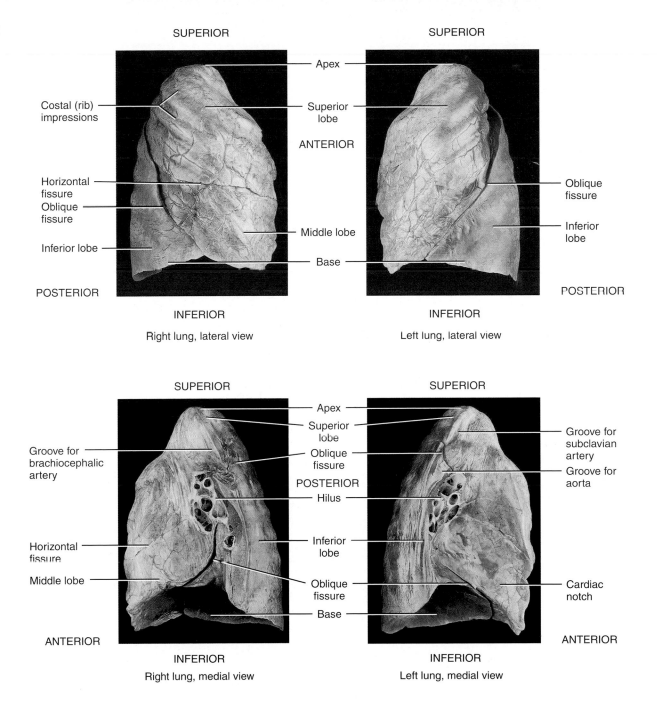

SUPERIOR

Apex

Costal (rib) impressions

Superior lobe

ANTERIOR

Horizontal fissure

Oblique fissure

Oblique fissure

Inferior lobe

Middle lobe

Inferior lobe

Base

POSTERIOR

POSTERIOR

INFERIOR

INFERIOR

Right lung, lateral view

Left lung, lateral view

SUPERIOR

SUPERIOR

Apex

Superior lobe

Groove for brachiocephalic artery

Oblique fissure

Groove for subclavian artery

POSTERIOR

Groove for aorta

Hilus

Horizontal fissure

Inferior lobe

Middle lobe

Oblique fissure

Cardiac notch

Base

ANTERIOR

ANTERIOR

INFERIOR

INFERIOR

Right lung, medial view

Left lung, medial view

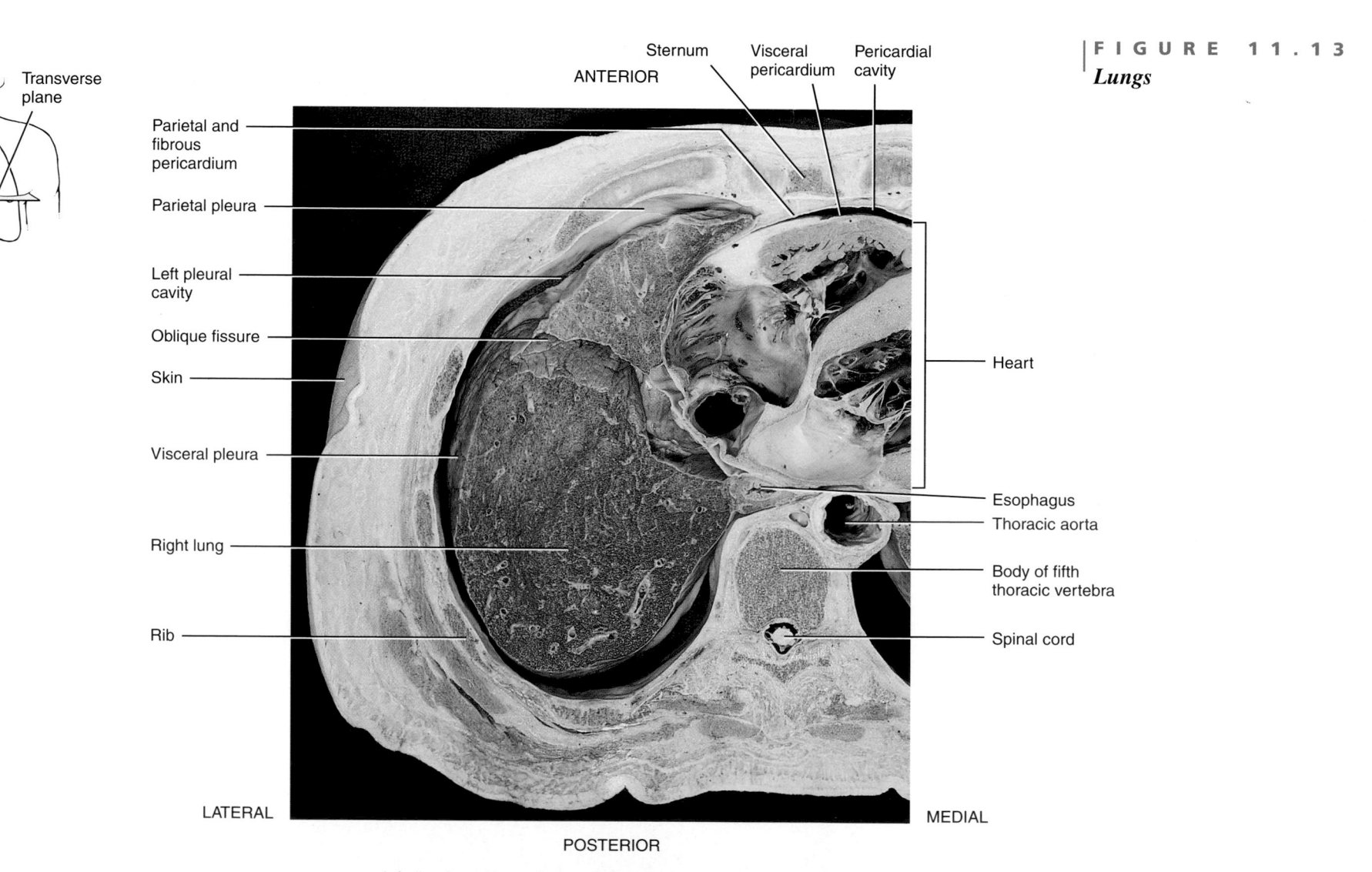

Transverse plane

View

Sternum

ANTERIOR

Visceral pericardium

Pericardial cavity

Parietal and fibrous pericardium

Parietal pleura

Left pleural cavity

Oblique fissure

Skin

Visceral pleura

Right lung

Rib

Heart

Esophagus

Thoracic aorta

Body of fifth thoracic vertebra

Spinal cord

LATERAL

MEDIAL

POSTERIOR

Inferior view of a transverse section through the thoracic cavity

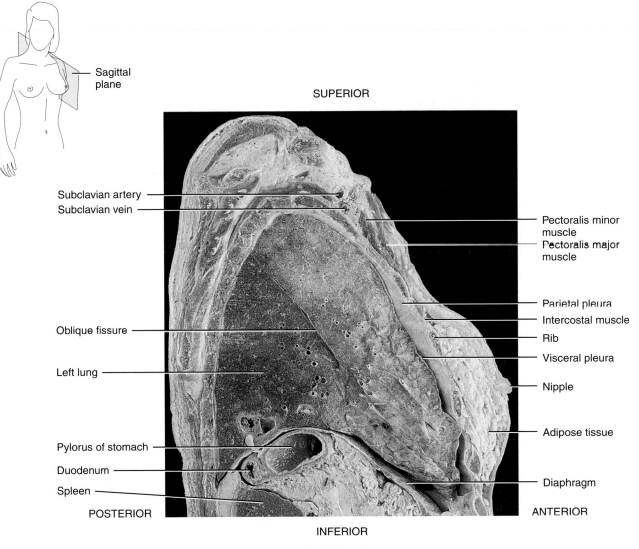

Sagittal
plane

SUPERIOR

Subclavian artery

Subclavian vein

Pectoralis minor
muscle

Pectoralis major
muscle

Parietal pleura

Intercostal muscle

Oblique fissure

Rib

Visceral pleura

Left lung

Nipple

Pylorus of stomach

Adipose tissue

Duodenum

Spleen

Diaphragm

POSTERIOR

ANTERIOR

INFERIOR

Sagittal section

FIGURE 11.14
Lungs

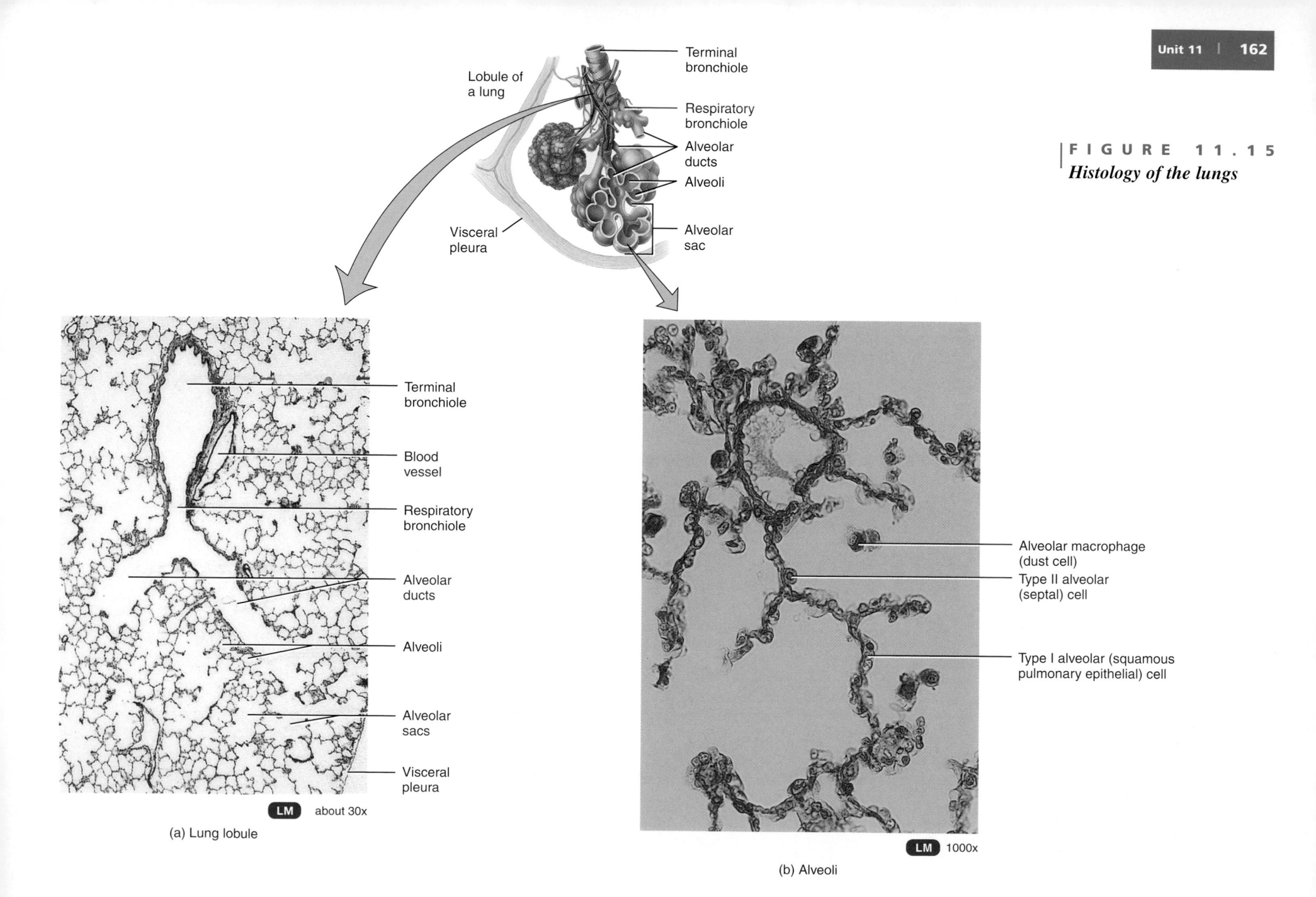

FIGURE 11.15
Histology of the lungs

Lobule of
a lung

Terminal
bronchiole

Respiratory
bronchiole

Alveolar
ducts

Alveoli

Alveolar
sac

Visceral
pleura

Terminal
bronchiole

Blood
vessel

Respiratory
bronchiole

Alveolar
ducts

Alveoli

Alveolar
sacs

Visceral
pleura

LM   about 30x

(a) Lung lobule

Alveolar macrophage
(dust cell)

Type II alveolar
(septal) cell

Type I alveolar (squamous
pulmonary epithelial) cell

LM   1000x

(b) Alveoli

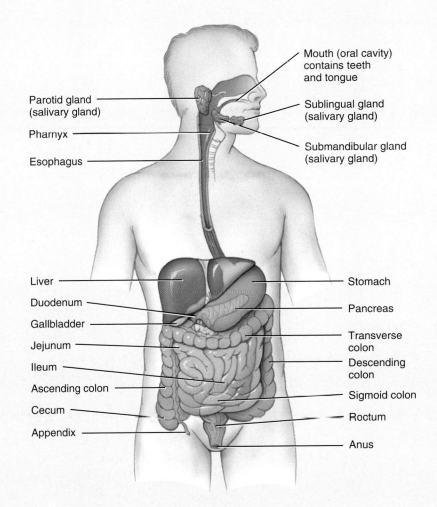

Parotid gland
(salivary gland)

Pharnyx

Esophagus

Mouth (oral cavity)
contains teeth
and tongue

Sublingual gland
(salivary gland)

Submandibular gland
(salivary gland)

Liver

Duodenum

Gallbladder

Jejunum

Ileum

Ascending colon

Cecum

Appendix

Stomach

Pancreas

Transverse
colon

Descending
colon

Sigmoid colon

Roctum

Anus

Right lateral view of head and neck and anterior view of trunk

FIGURE 12.1
Digestive system

SUPERIOR

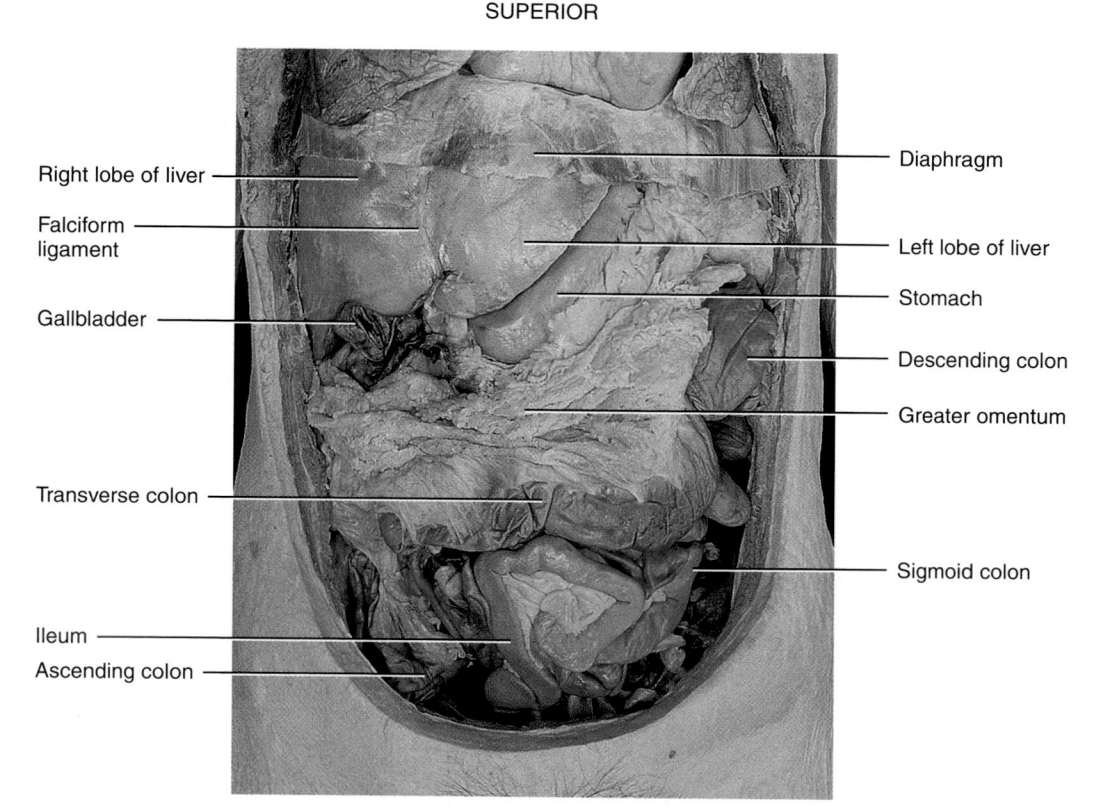

Right lobe of liver

Falciform ligament

Gallbladder

Transverse colon

Ileum

Ascending colon

Diaphragm

Left lobe of liver

Stomach

Descending colon

Greater omentum

Sigmoid colon

INFERIOR

Anterior view

FIGURE 12.2

Digestive organs of the abdominopelvic cavity

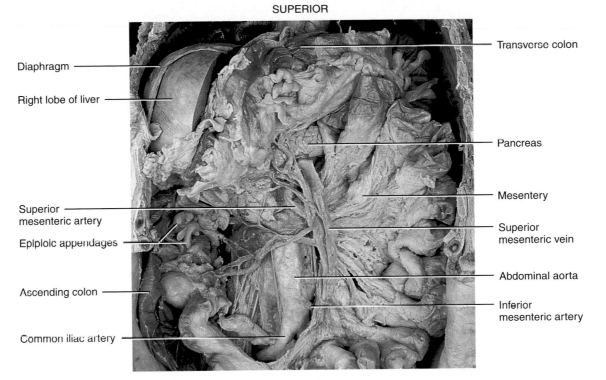

SUPERIOR

Diaphragm

Right lobe of liver

Superior
mesenteric artery

Epiploic appendages

Ascending colon

Common iliac artery

Transverse colon

Pancreas

Mesentery

Superior
mesenteric vein

Abdominal aorta

Inferior
mesenteric artery

INFERIOR

Anterior view

F I G U R E   1 2 . 3
Digestive organs of the abdominopelvic cavity

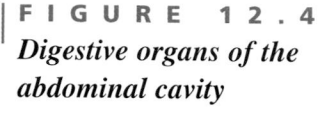

FIGURE 12.4
Digestive organs of the abdominal cavity

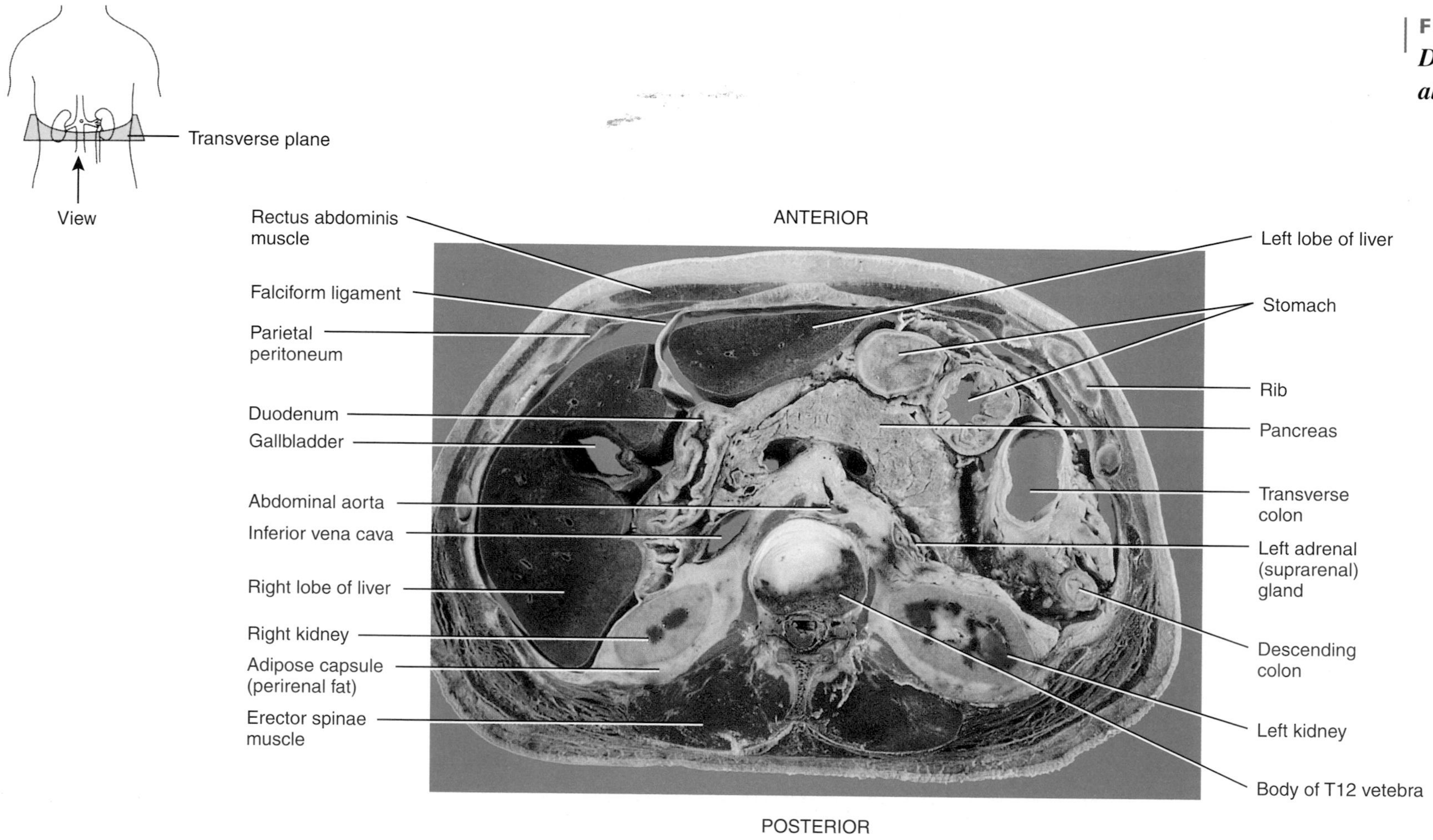

Transverse plane

View

ANTERIOR

Rectus abdominis muscle

Falciform ligament

Parietal peritoneum

Duodenum

Gallbladder

Abdominal aorta

Inferior vena cava

Right lobe of liver

Right kidney

Adipose capsule (perirenal fat)

Erector spinae muscle

Left lobe of liver

Stomach

Rib

Pancreas

Transverse colon

Left adrenal (suprarenal) gland

Descending colon

Left kidney

Body of T12 vetebra

POSTERIOR

Inferior view of a transverse section through the abdomen

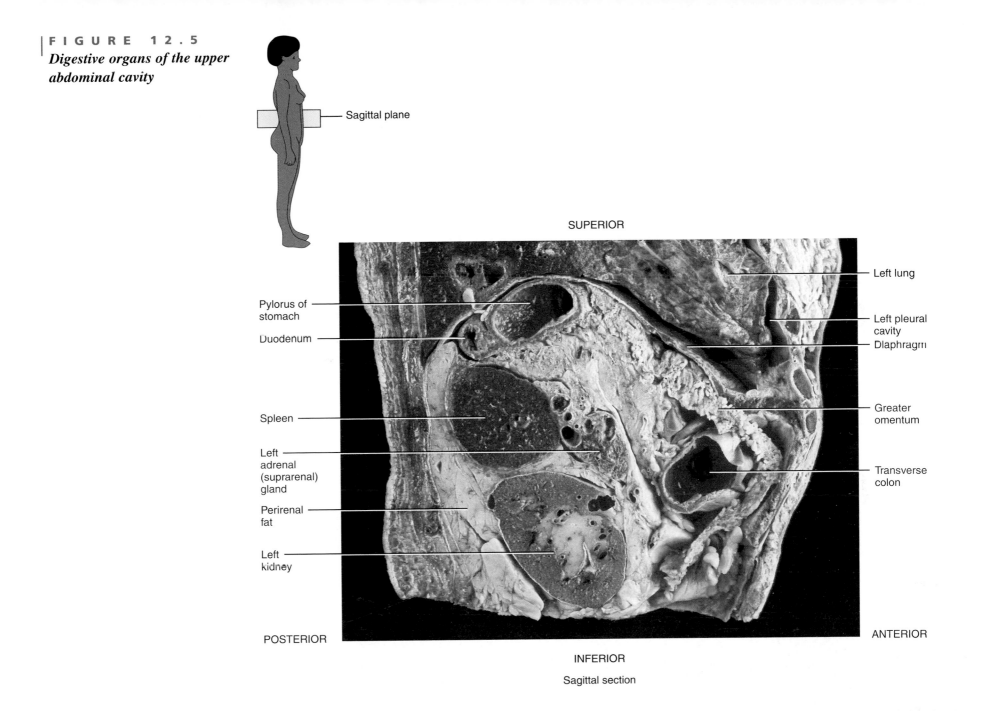

FIGURE 12.5

Digestive organs of the upper abdominal cavity

Sagittal plane

SUPERIOR

Left lung

Pylorus of stomach

Left pleural cavity

Duodenum

Diaphragm

Spleen

Greater omentum

Left adrenal (suprarenal) gland

Transverse colon

Perirenal fat

Left kidney

POSTERIOR

ANTERIOR

INFERIOR

Sagittal section

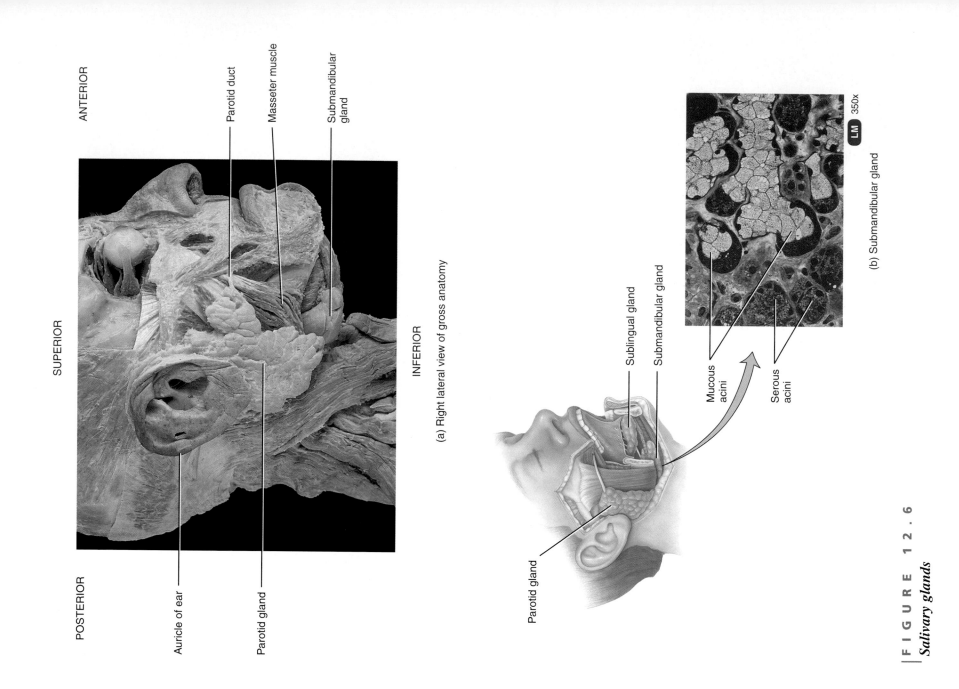

ANTERIOR

Parotid duct

Masseter muscle

Submandibular gland

SUPERIOR

INFERIOR

POSTERIOR

Auricle of ear

Parotid gland

(a) Right lateral view of gross anatomy

Sublingual gland

Submandibular gland

Mucous acini

Serous acini

LM 350x

(b) Submandibular gland

Parotid gland

FIGURE 12.6

Salivary glands

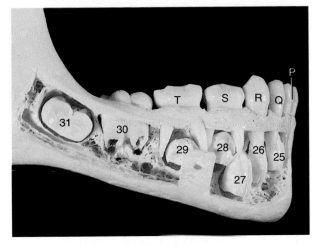

Right lateral view

Deciduous teeth:
P - Central incisor
Q - Lateral incisor
R - Cuspid (canine)
S - First molar (bicuspid)
T - Second molar

Permanent teeth:
25 - Central incisor
26 - Lateral incisor
27 - Cuspid (canine)
28 - First premolar (bicuspid)
29 - Second premolar
30 - First molar
31 - Second molar

(a) Mandible of a six year old child showing erupted deciduous teeth and unerupted permanent teeth

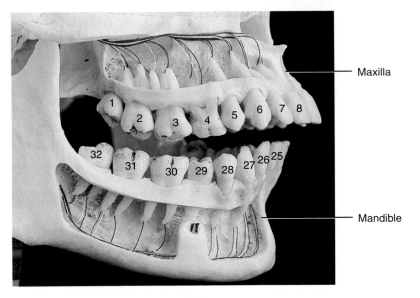

Maxilla

Mandible

Right lateral view

8, 25 - Central incisor
7, 26 - Lateral incisor
6, 27 - Cuspid (canine)
5, 28 - First premolar (bicuspid)
4, 29 - Second premolar
3, 30 - First molar
2, 31 - Second molar
1, 32 - Third molar (wisdom tooth)

(b) Mandible (and maxilla) showing permanent teeth and blood and nerve supply to them

FIGURE 12.7 *Teeth*

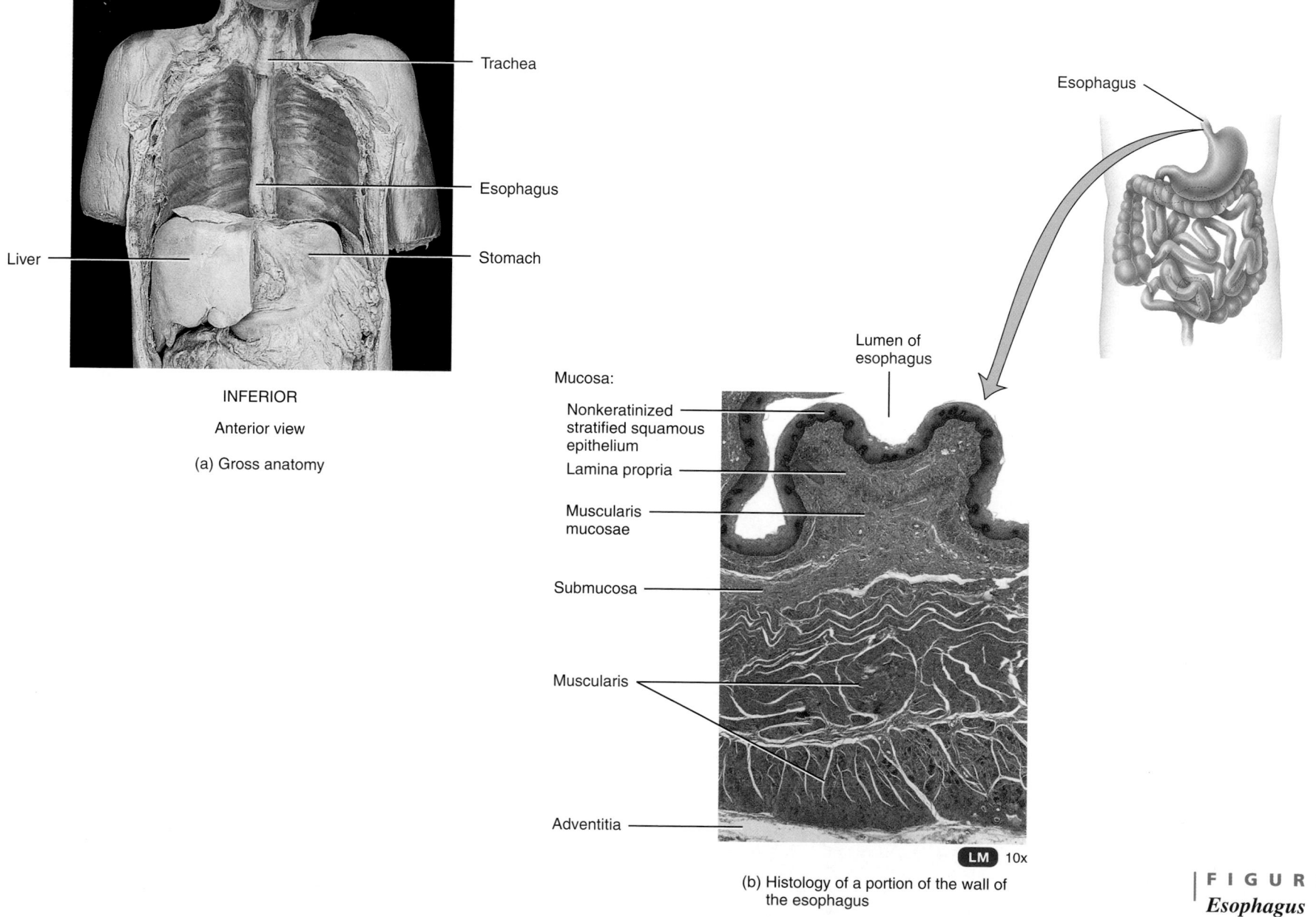

SUPERIOR

Trachea

Esophagus

Liver

Stomach

INFERIOR

Anterior view

(a) Gross anatomy

Esophagus

Lumen of
esophagus

Mucosa:

Nonkeratinized
stratified squamous
epithelium

Lamina propria

Muscularis
mucosae

Submucosa

Muscularis

Adventitia

LM 10x

(b) Histology of a portion of the wall of
the esophagus

FIGURE 12.8
Esophagus

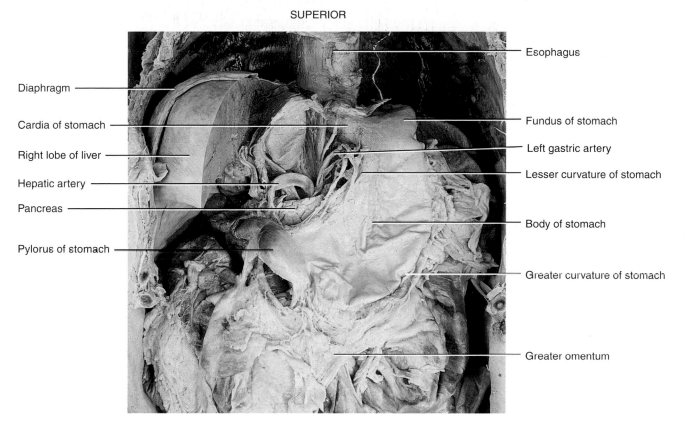

Esophagus

Diaphragm

Cardia of stomach

Fundus of stomach

Right lobe of liver

Left gastric artery

Hepatic artery

Lesser curvature of stomach

Pancreas

Body of stomach

Pylorus of stomach

Greater curvature of stomach

Greater omentum

INFERIOR

(a) Anterior view of external anatomy

F I G U R E   1 2 . 9
Stomach

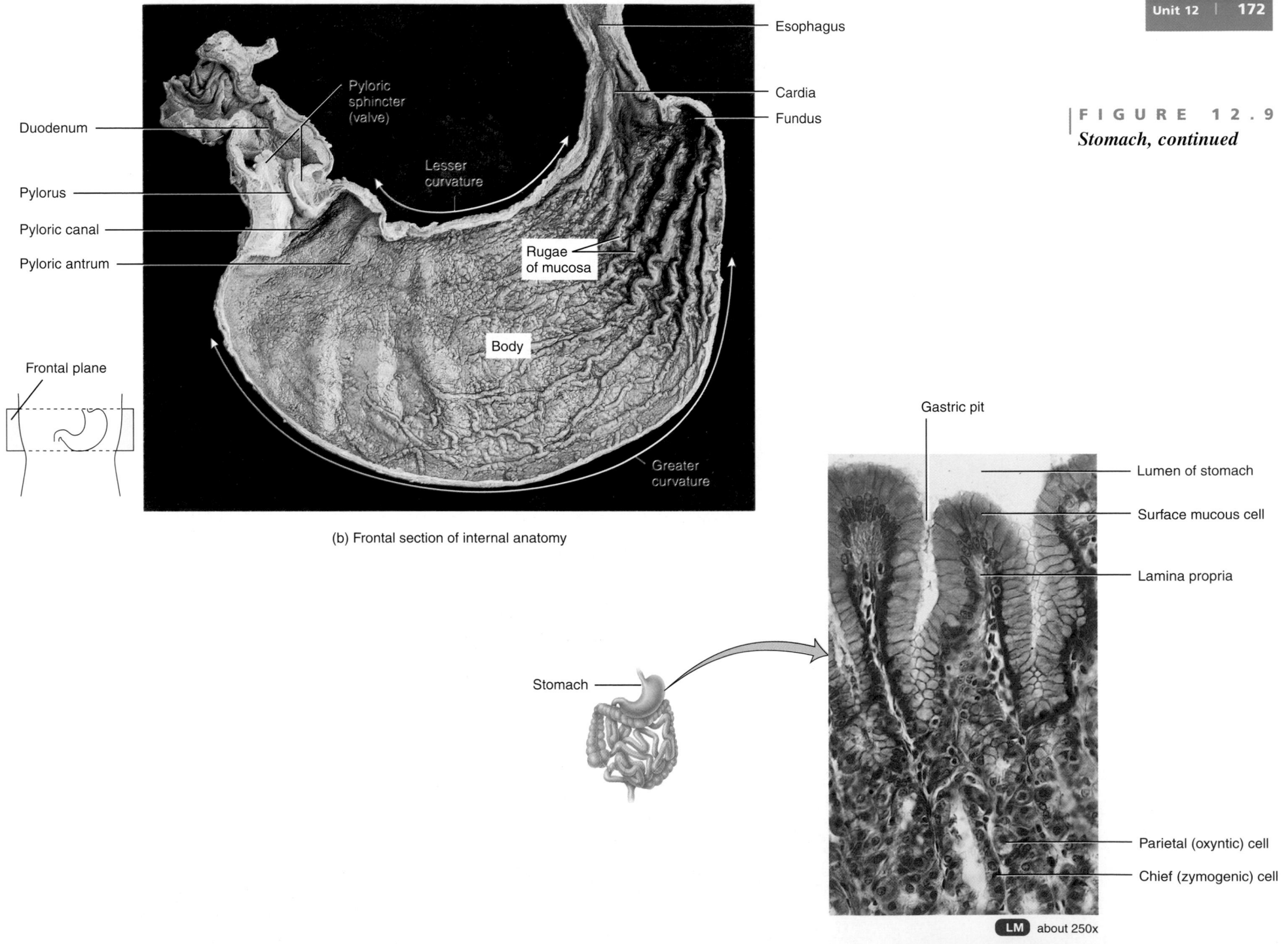

Esophagus

Cardia

Fundus

FIGURE 12.9
Stomach, continued

Pyloric
sphincter
(valve)

Duodenum

Lesser
curvature

Pylorus

Pyloric canal

Pyloric antrum

Rugae
of mucosa

Body

Frontal plane

Greater
curvature

(b) Frontal section of internal anatomy

Gastric pit

Lumen of stomach

Surface mucous cell

Lamina propria

Stomach

Parietal (oxyntic) cell

Chief (zymogenic) cell

LM about 250x

(c) Histology of fundic mucosa

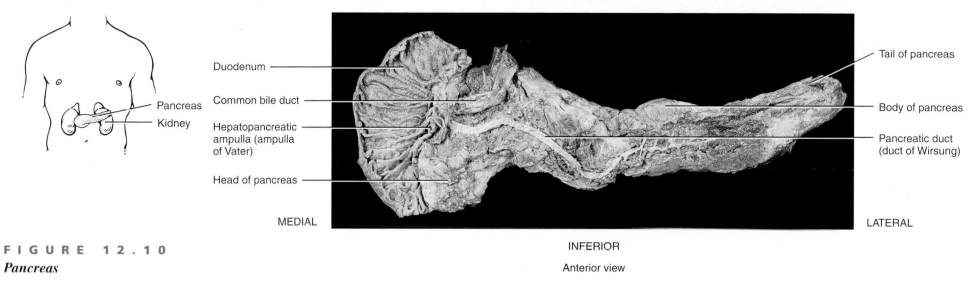

SUPERIOR

Duodenum

Common bile duct

Hepatopancreatic
ampulla (ampulla
of Vater)

Head of pancreas

Tail of pancreas

Body of pancreas

Pancreatic duct
(duct of Wirsung)

MEDIAL

LATERAL

INFERIOR

Anterior view

Pancreas

Kidney

FIGURE 12.10

Pancreas

FIGURE 12.11

Liver and gallbladder

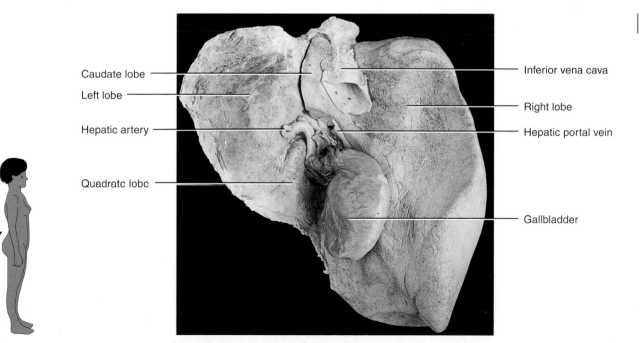

Caudate lobe

Left lobe

Hepatic artery

Quadrate lobe

View

Inferior vena cava

Right lobe

Hepatic portal vein

Gallbladder

(a) Posteroinferior view of gross anatomy

Transverse plane

View

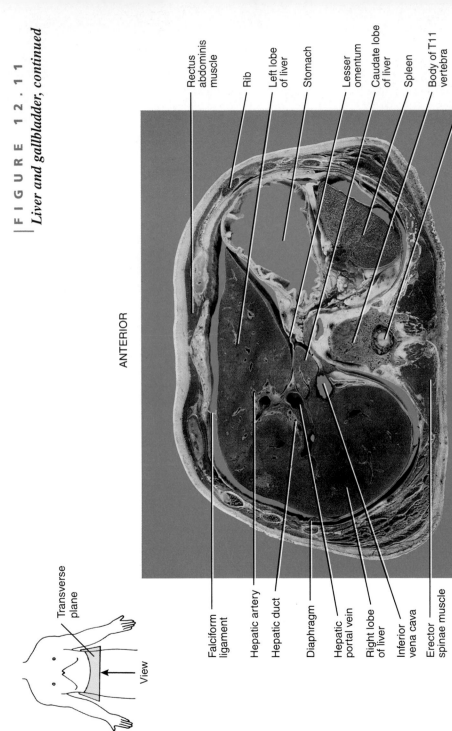

ANTERIOR

Rectus abdominis muscle

Rib

Left lobe of liver

Stomach

Lesser omentum

Caudate lobe of liver

Spleen

Body of T11 vertebra

Spinal cord

Falciform ligament

Hepatic artery

Hepatic duct

Diaphragm

Hepatic portal vein

Right lobe of liver

Inferior vena cava

Erector spinae muscle

POSTERIOR

(b) Inferior view of a transverse section of the abdomen

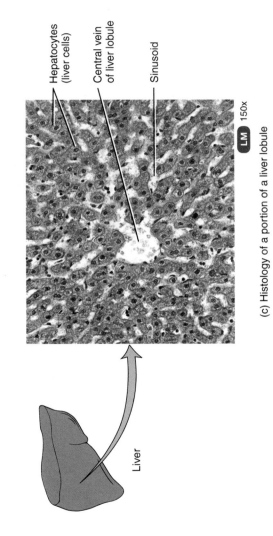

Hepatocytes (liver cells)

Central vein of liver lobule

Sinusoid

LM 150x

Liver

(c) Histology of a portion of a liver lobule

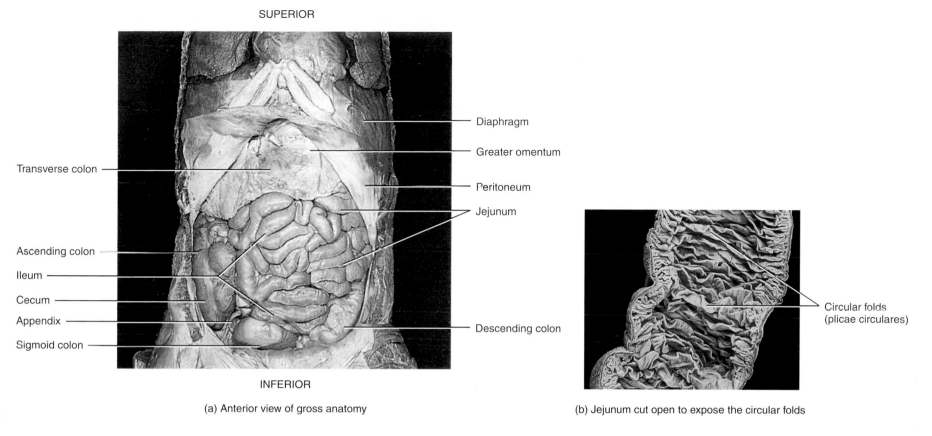

SUPERIOR

Transverse colon

Ascending colon

Ileum

Cecum

Appendix

Sigmoid colon

INFERIOR

Diaphragm

Greater omentum

Peritoneum

Jejunum

Descending colon

(a) Anterior view of gross anatomy

Circular folds
(plicae circulares)

(b) Jejunum cut open to expose the circular folds

FIGURE 12.12
Small intestine

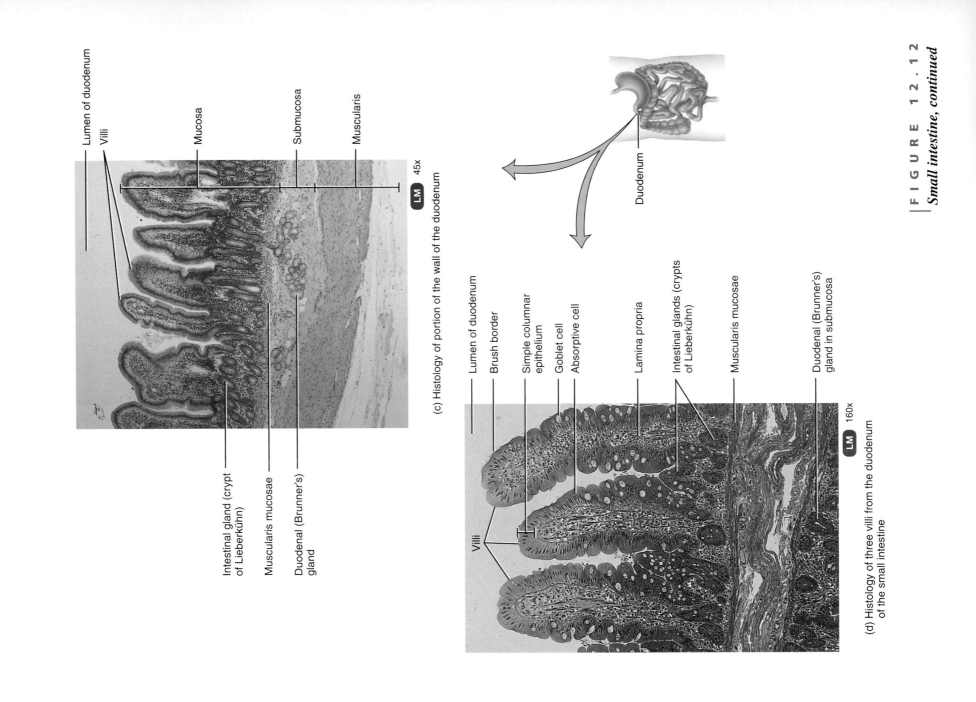

Lumen of duodenum

Villi

Mucosa

Submucosa

Muscularis

LM 45x

Intestinal gland (crypt
of Lieberkühn)

Muscularis mucosae

Duodenal (Brunner's)
gland

(c) Histology of portion of the wall of the duodenum

Duodenum

Lumen of duodenum

Brush border

Simple columnar
epithelium

Goblet cell

Absorptive cell

Lamina propria

Intestinal glands (crypts
of Lieberkühn)

Muscularis mucosae

Duodenal (Brunner's)
gland in submucosa

Villi

LM 160x

(d) Histology of three villi from the duodenum
of the small intestine

F I G U R E   1 2 . 1 2
Small intestine, continued

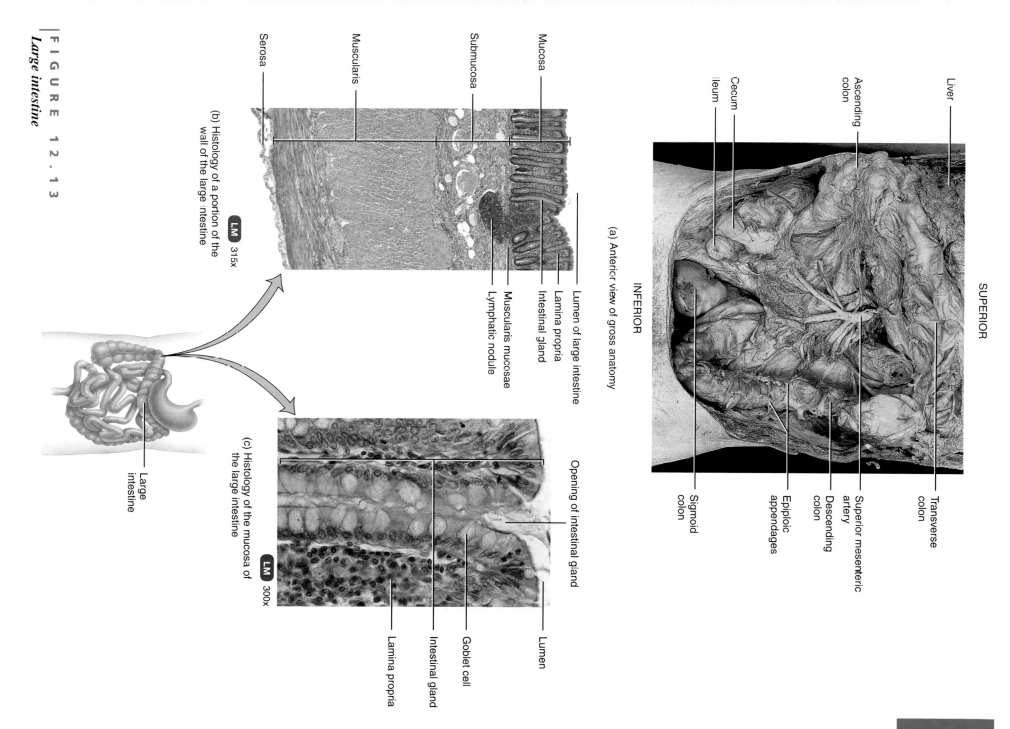

F I G U R E   1 2 . 1 3

Large intestine

(a) Anterior view of gross anatomy

SUPERIOR

INFERIOR

Liver

Transverse colon

Ascending colon

Superior mesenteric artery

Descending colon

Cecum

Ileum

Epiploic appendages

Sigmoid colon

(b) Histology of a portion of the wall of the large intestine

LM 315x

Serosa

Muscularis

Submucosa

Mucosa

Lumen of large intestine

Lamina propria

Intestinal gland

Muscularis mucosae

Lymphatic nodule

Large intestine

(c) Histology of the mucosa of the large intestine

LM 300x

Opening of intestinal gland

Lumen

Goblet cell

Intestinal gland

Lamina propria

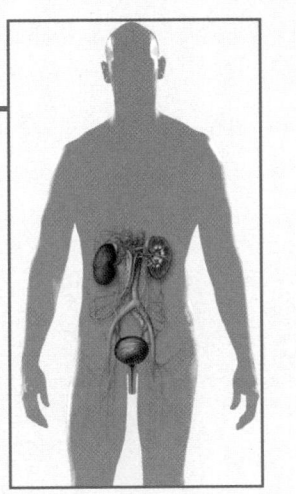

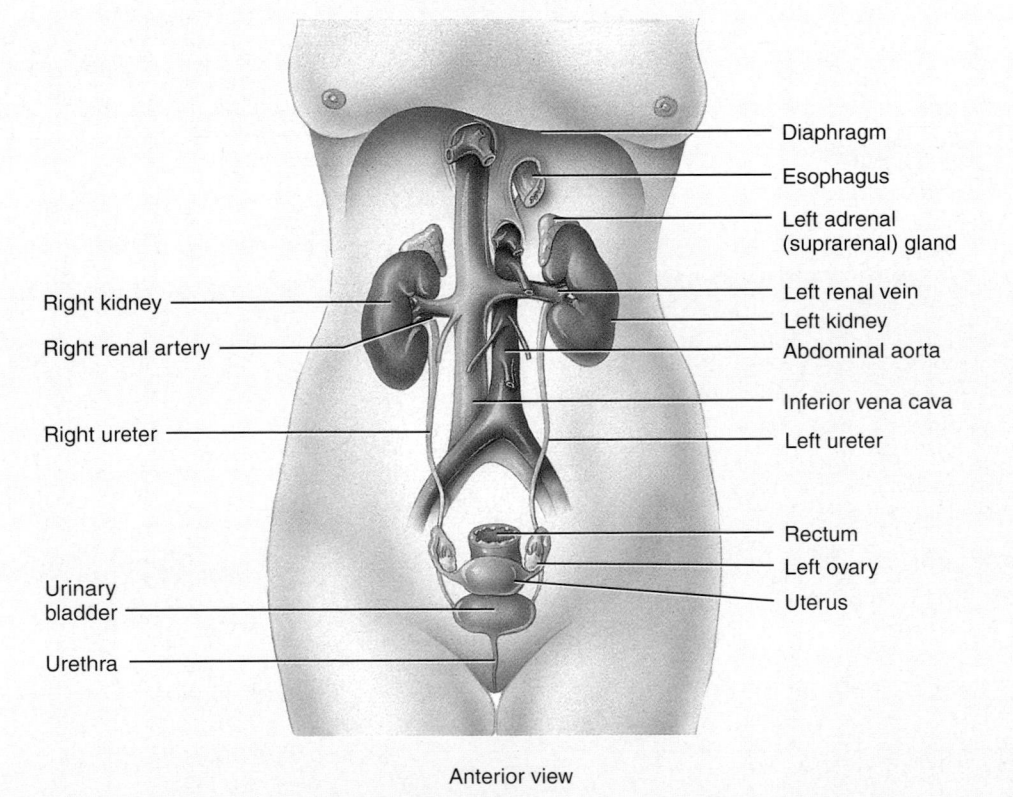

Diaphragm

Esophagus

Left adrenal (suprarenal) gland

Right kidney

Left renal vein

Right renal artery

Left kidney

Abdominal aorta

Inferior vena cava

Right ureter

Left ureter

Rectum

Left ovary

Urinary bladder

Uterus

Urethra

Anterior view

FIGURE 13.1
Urinary system

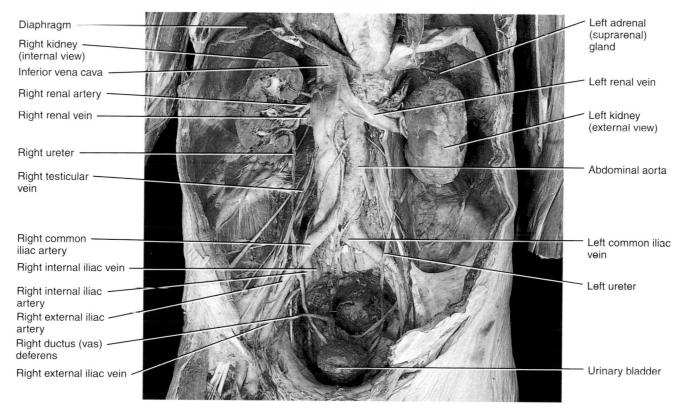

Diaphragm

Right kidney
(internal view)

Inferior vena cava

Right renal artery

Right renal vein

Right ureter

Right testicular
vein

Right common
iliac artery

Right internal iliac vein

Right internal iliac
artery

Right external iliac
artery

Right ductus (vas)
deferens

Right external iliac vein

Left adrenal
(suprarenal)
gland

Left renal vein

Left kidney
(external view)

Abdominal aorta

Left common iliac
vein

Left ureter

Urinary bladder

INFERIOR

Anterior view

FIGURE 13.2
Urinary organs

SUPERIOR

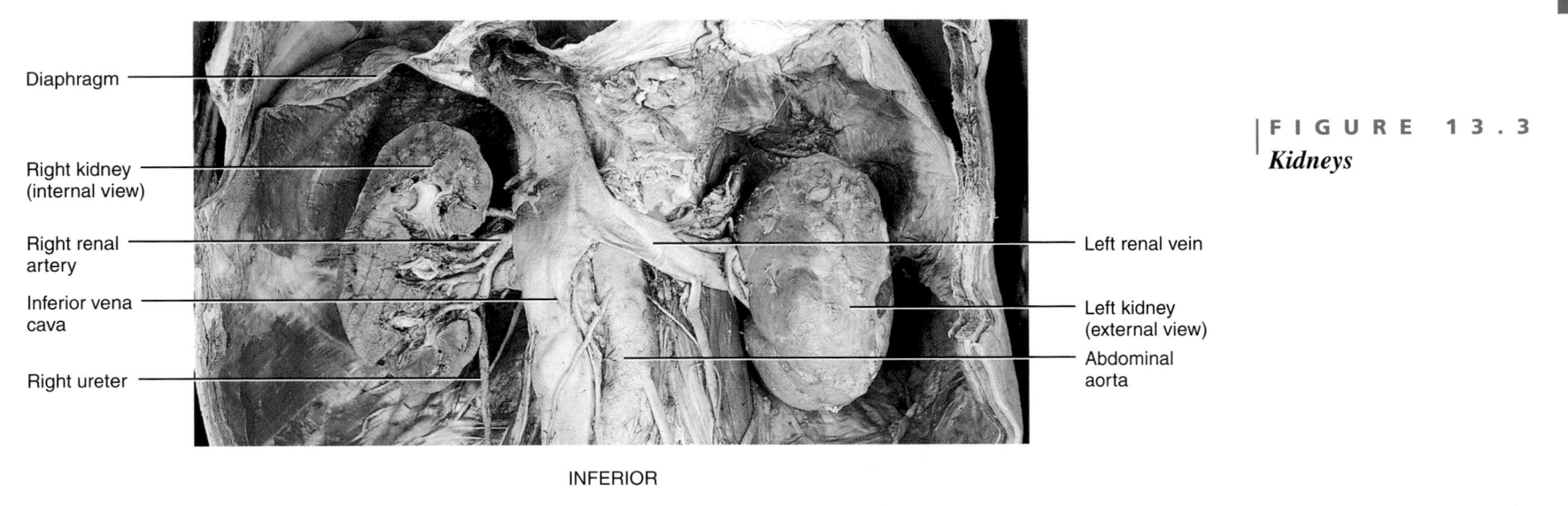

Diaphragm

Right kidney
(internal view)

Right renal
artery

Inferior vena
cava

Right ureter

Left renal vein

Left kidney
(external view)

Abdominal
aorta

INFERIOR

Anterior view

FIGURE 13.3
Kidneys

SUPERIOR

Adrenal (suprarenal) gland

Inferior vena cava

Suprarenal arteries

Renal artery

Renal vein

Kidney

Ureter

FIGURE 13.4
Right kidney, external aspect

LATERAL

MEDIAL

INFERIOR

Anterior view

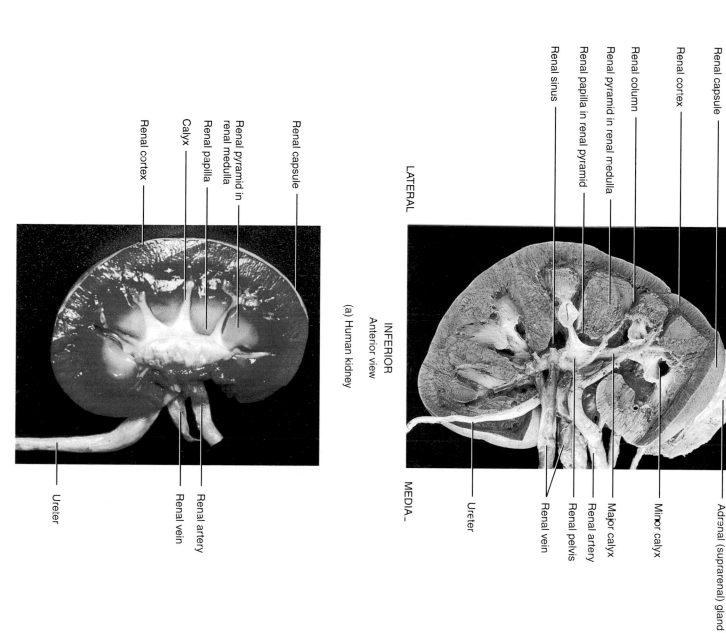

Frontal
plane

Renal sinus

Renal papilla in renal pyramid

Renal pyramid in renal medulla

Renal column

Renal cortex

Renal capsule

SUPERIOR

LATERAL

INFERIOR
Anterior view

(a) Human kidney

MEDIA_

Ureter

Renal vein

Renal pelvis

Renal artery

Major calyx

Minor calyx

Adrenal (suprarenal) gland

Renal cortex

Calyx

Renal papilla

Renal pyramid in
renal medulla

Renal capsule

Ureter

Renal vein

Renal artery

(b) Sheep kidney

F I G U R E 1 3 . 5
*Right kidney, internal aspect. A transverse section of the
kidneys is illustrated in Fig. 12.4.*

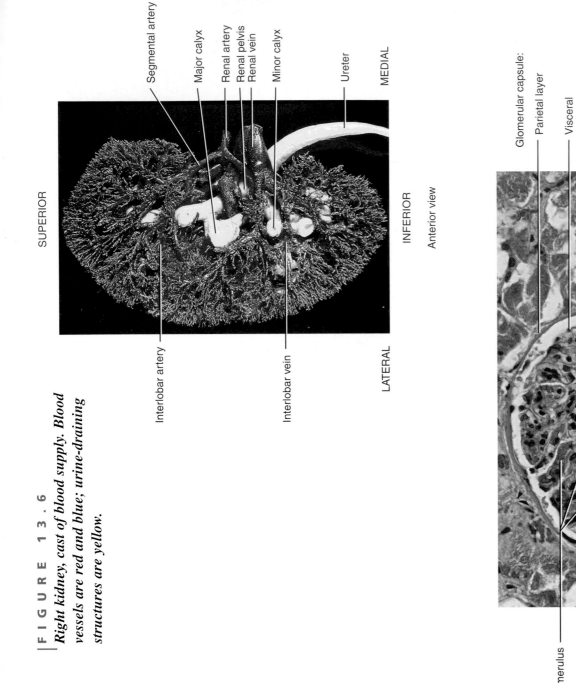

FIGURE 13.6
Right kidney, cast of blood supply. Blood vessels are red and blue; urine-draining structures are yellow.

SUPERIOR

Segmental artery

Major calyx

Renal artery
Renal pelvis
Renal vein

Minor calyx

Ureter

MEDIAL

Interlobar artery

Interlobar vein

INFERIOR

Anterior view

LATERAL

FIGURE 13.7
Histology of the kidney

Glomerular capsule:

Parietal layer

Visceral layer

Afferent arteriole

Juxtaglomerular cell

Ascending limb of loop of Henle

Macula densa cell

Efferent arteriole

Proximal convoluted tubule

LM 1380x

Renal corpuscle and surrounding structures

Kidney

Glomerulus

Podocytes of visceral layer of glomerular capsule

Capsular space

Simple squamous epithelial cells

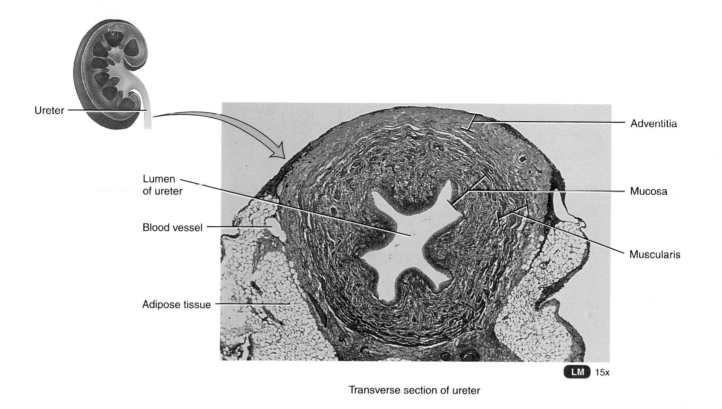

Ureter

Adventitia

Lumen
of ureter

Mucosa

Blood vessel

Muscularis

Adipose tissue

LM 15x

Transverse section of ureter

FIGURE 13.8
Histology of the ureter

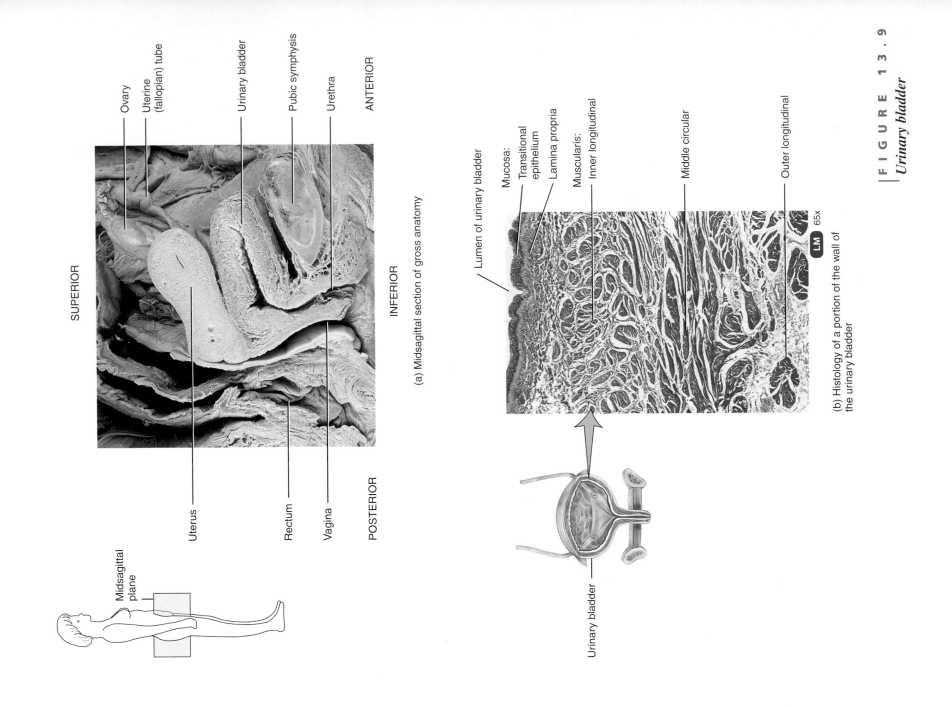

SUPERIOR

Ovary

Uterine (fallopian) tube

Urinary bladder

Pubic symphysis

Urethra

ANTERIOR

Uterus

Rectum

Vagina

POSTERIOR

INFERIOR

(a) Midsagittal section of gross anatomy

Midsagittal plane

Lumen of urinary bladder

Mucosa:
Transitional epithelium
Lamina propria

Muscularis:
Inner longitudinal

Middle circular

Outer longitudinal

LM 65x

(b) Histology of a portion of the wall of the urinary bladder

Urinary bladder

FIGURE 13.9
Urinary bladder

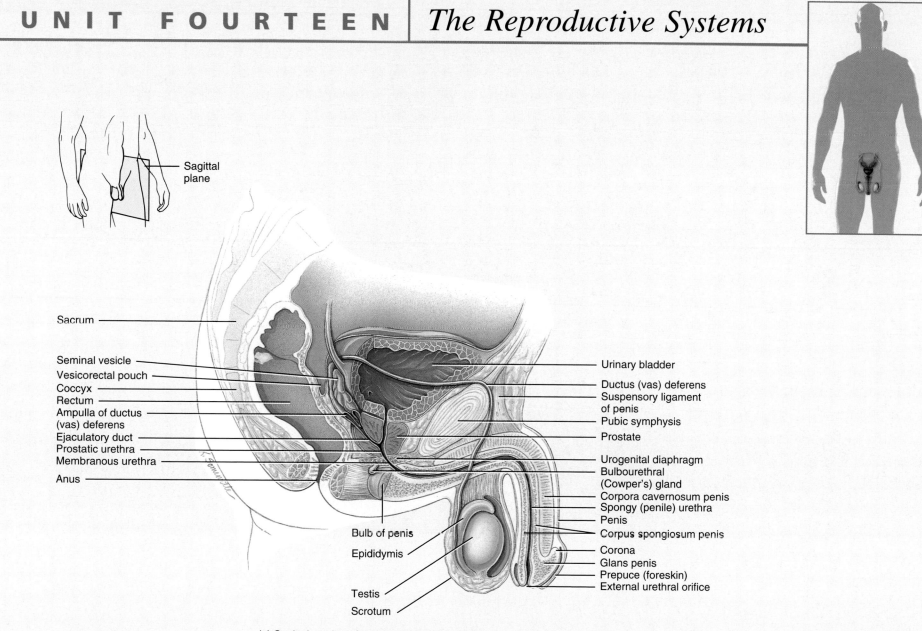

Sagittal plane

Sacrum

Seminal vesicle
Vesicorectal pouch
Coccyx
Rectum
Ampulla of ductus (vas) deferens
Ejaculatory duct
Prostatic urethra
Membranous urethra

Anus

Urinary bladder

Ductus (vas) deferens
Suspensory ligament of penis

Pubic symphysis

Prostate

Urogenital diaphragm
Bulbourethral (Cowper's) gland
Corpora cavernosum penis
Spongy (penile) urethra
Penis
Corpus spongiosum penis

Corona
Glans penis
Prepuce (foreskin)
External urethral orifice

Bulb of penis
Epididymis

Testis
Scrotum

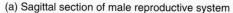

(a) Sagittal section of male reproductive system

FIGURE 14.1
Reproductive systems

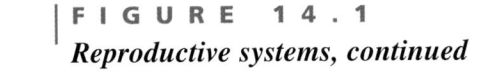

FIGURE 14.1
Reproductive systems, continued

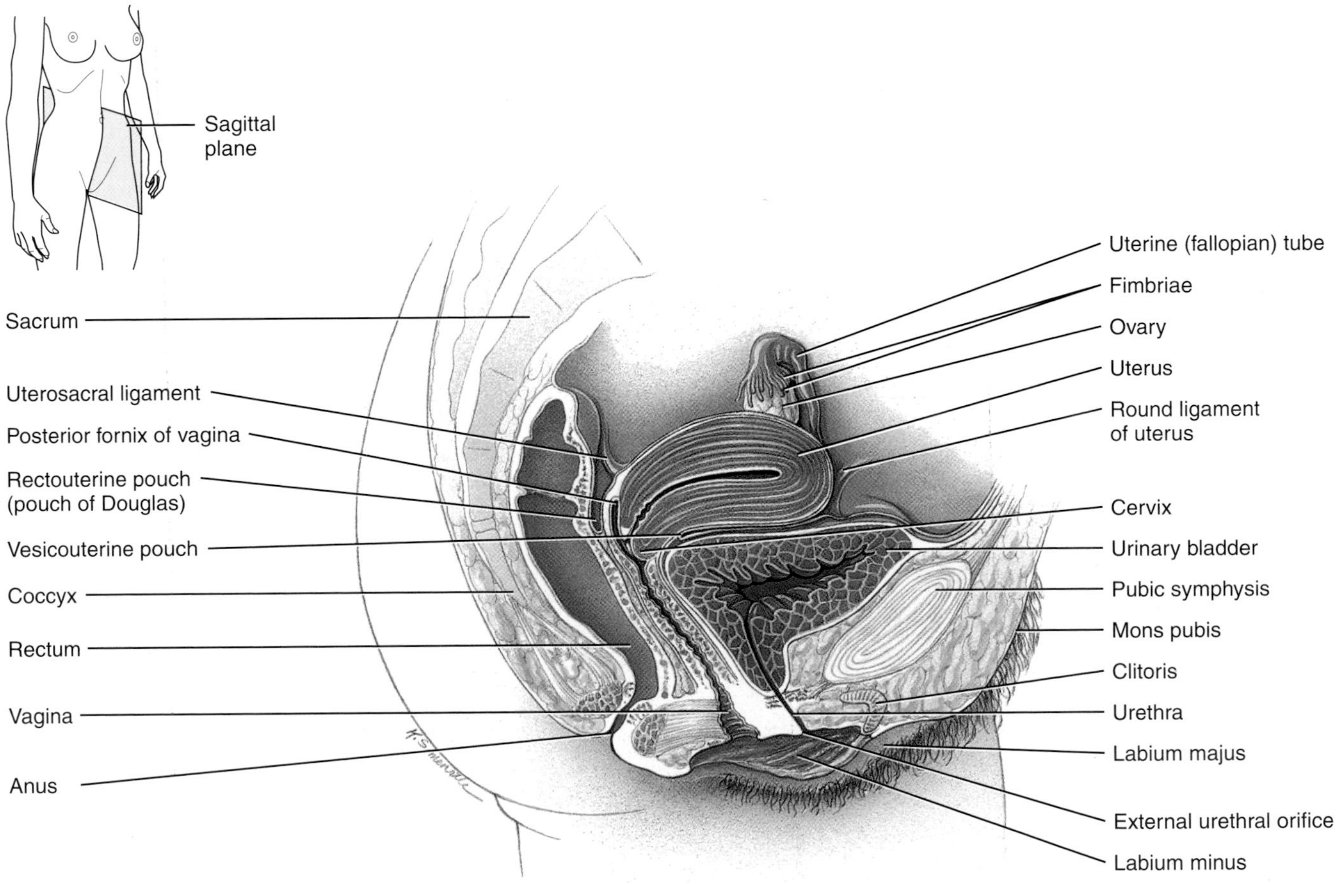

Sagittal plane

Uterine (fallopian) tube

Fimbriae

Sacrum

Ovary

Uterus

Uterosacral ligament

Round ligament of uterus

Posterior fornix of vagina

Rectouterine pouch (pouch of Douglas)

Cervix

Vesicouterine pouch

Urinary bladder

Coccyx

Pubic symphysis

Rectum

Mons pubis

Vagina

Clitoris

Urethra

Anus

Labium majus

External urethral orifice

Labium minus

(b) Sagittal section of female reproductive system

SUPERIOR

POSTERIOR

ANTERIOR

Ureter

Urinary bladder
(opened)

Pubic symphysis

Ductus (vas)
deferens

Corpora
cavernosum penis

Right ureter

Corpus
spongiosum penis

Seminal vesicle
(sectioned)

Spongy (penile)
urethra

Ampulla of ductus
(vas) deferens

Corona

Prostatic urethra

Glans penis

Ejaculatory duct

Prostate

Bulbospongiosus
muscle

Crus of penis
covered by
ischiocavernosus
muscle

Bulb of penis

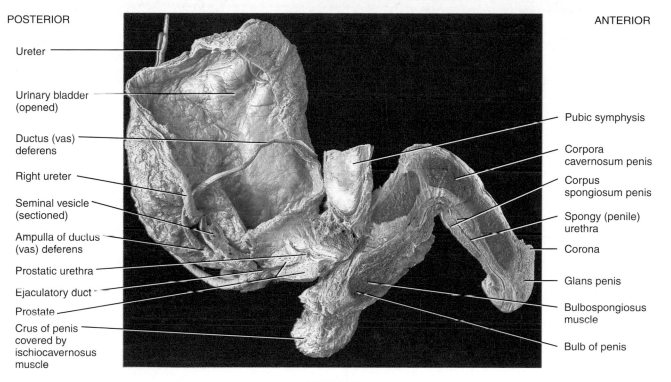

INFERIOR

Sagittal dissection

F I G U R E   1 4 . 2
Male reproductive organs

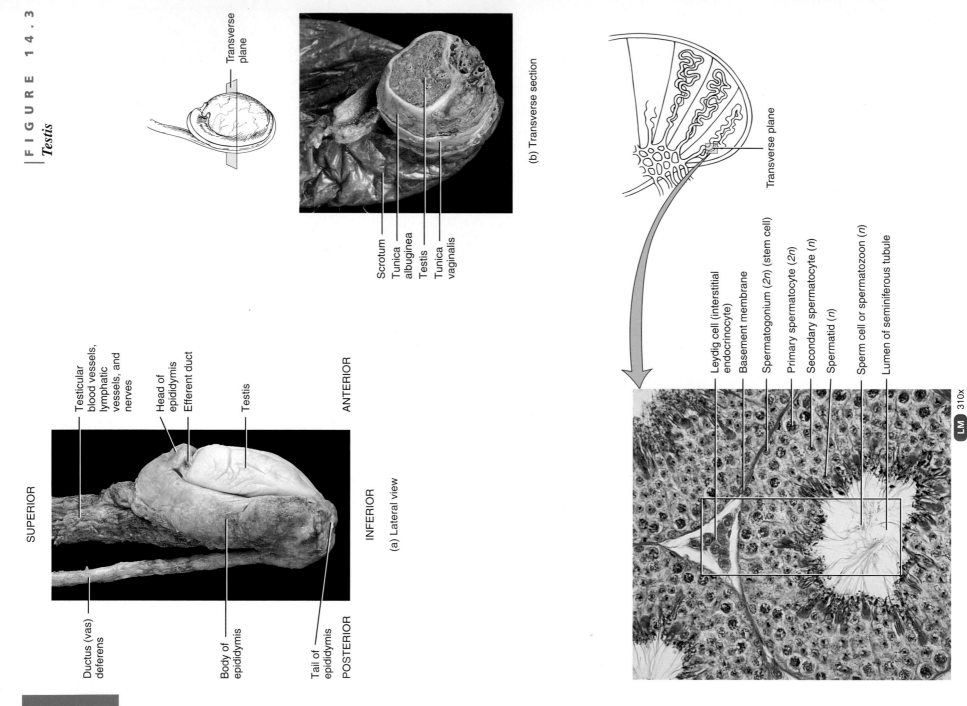

FIGURE 14.3
Testis

Transverse plane

SUPERIOR

Testicular blood vessels, lymphatic vessels, and nerves

Head of epididymis

Efferent duct

Testis

Ductus (vas) deferens

Body of epididymis

Tail of epididymis

POSTERIOR

ANTERIOR

INFERIOR

(a) Lateral view

Scrotum
Tunica albuginea
Testis
Tunica vaginalis

(b) Transverse section

Transverse plane

Leydig cell (interstitial endocrinocyte)
Basement membrane
Spermatogonium (2n) (stem cell)
Primary spermatocyte (2n)
Secondary spermatocyte (n)
Spermatid (n)
Sperm cell or spermatozoon (n)
Lumen of seminiferous tubule

LM 310x

(c) Histology of transverse section of several seminiferous tubules

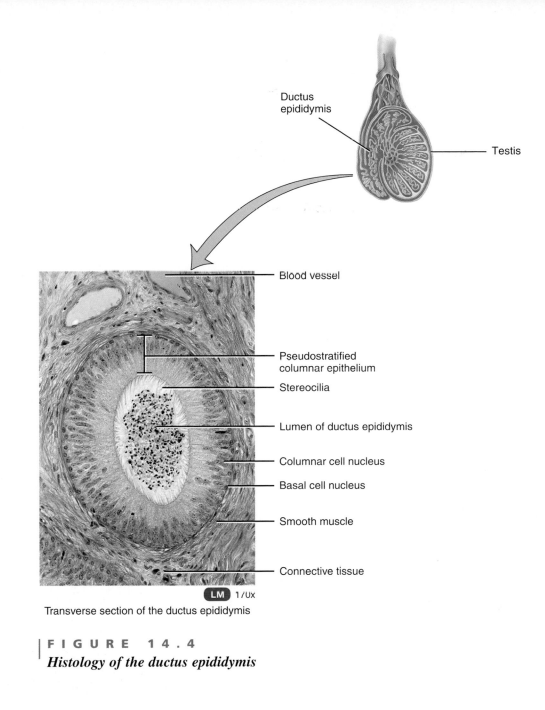

Ductus
epididymis

Testis

Blood vessel

Pseudostratified
columnar epithelium

Stereocilia

Lumen of ductus epididymis

Columnar cell nucleus

Basal cell nucleus

Smooth muscle

Connective tissue

LM 1/0x

Transverse section of the ductus epididymis

FIGURE 14.4
Histology of the ductus epididymis

FIGURE 14.5
Histology of the ductus (vas) deferens

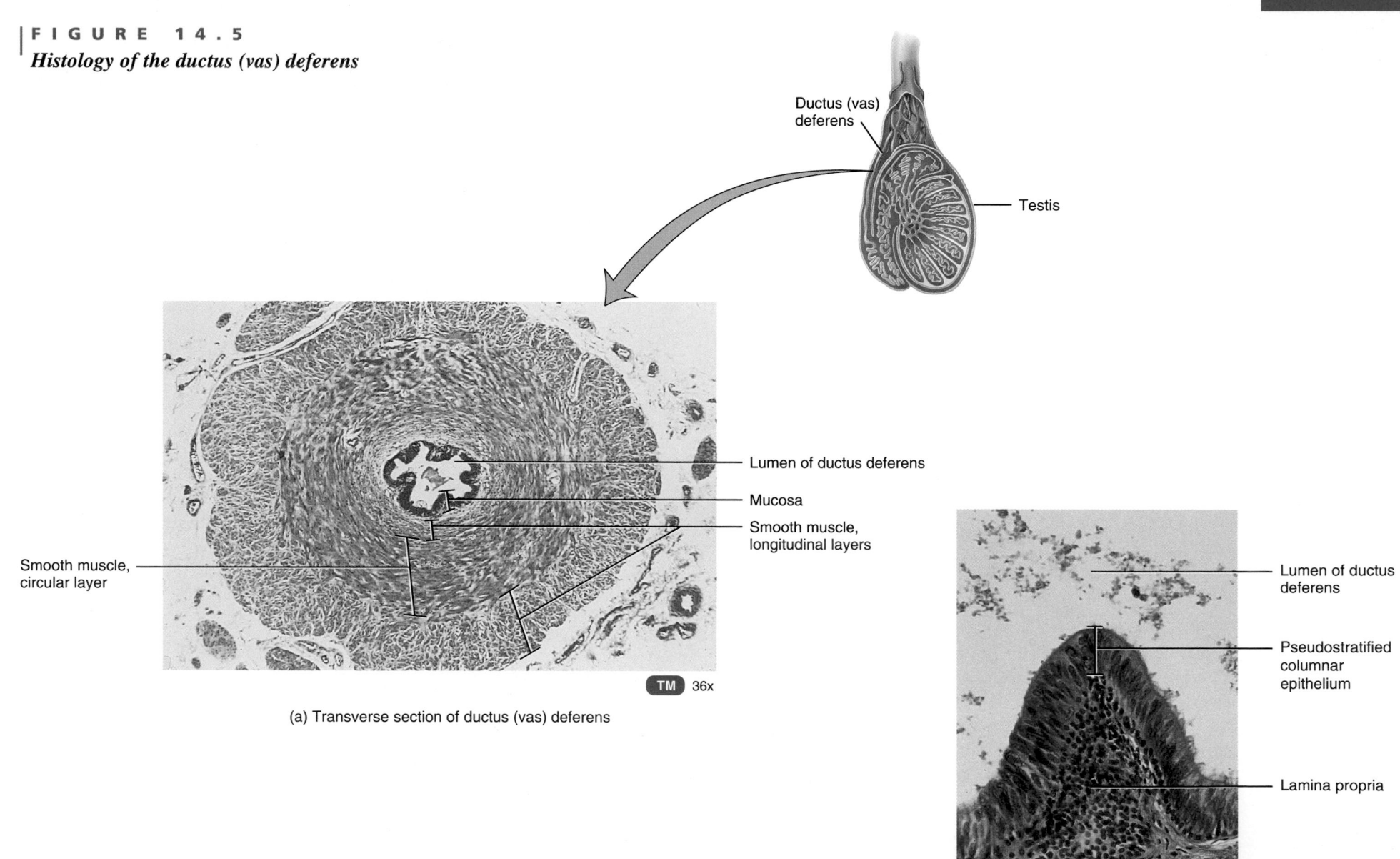

Ductus (vas) deferens

Testis

Lumen of ductus deferens

Mucosa

Smooth muscle, longitudinal layers

Smooth muscle, circular layer

TM 36x

(a) Transverse section of ductus (vas) deferens

Lumen of ductus deferens

Pseudostratified columnar epithelium

Lamina propria

Smooth muscle, longitudinal layer

LM 170x

(b) Transverse section of mucosa of ductus (vas) deferens

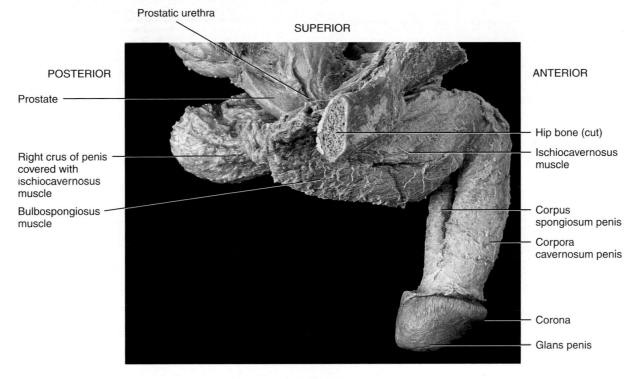

Prostatic urethra

SUPERIOR

POSTERIOR

ANTERIOR

Prostate

Hip bone (cut)

Ischiocavernosus muscle

Right crus of penis covered with ischiocavernosus muscle

Bulbospongiosus muscle

Corpus spongiosum penis

Corpora cavernosum penis

Corona

Glans penis

INFERIOR

Right lateral view

FIGURE 14.6
Penis

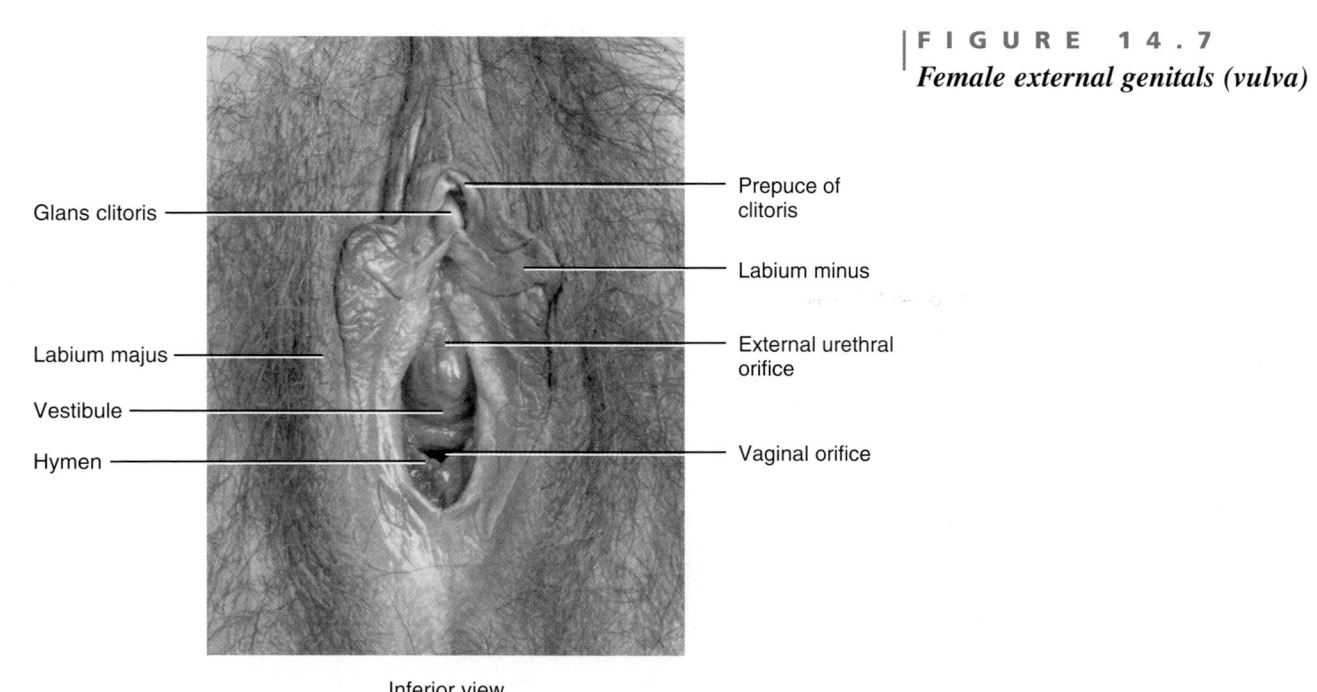

Glans clitoris

Prepuce of
clitoris

Labium minus

Labium majus

External urethral
orifice

Vestibule

Hymen

Vaginal orifice

Inferior view

FIGURE 14.7
Female external genitals (vulva)

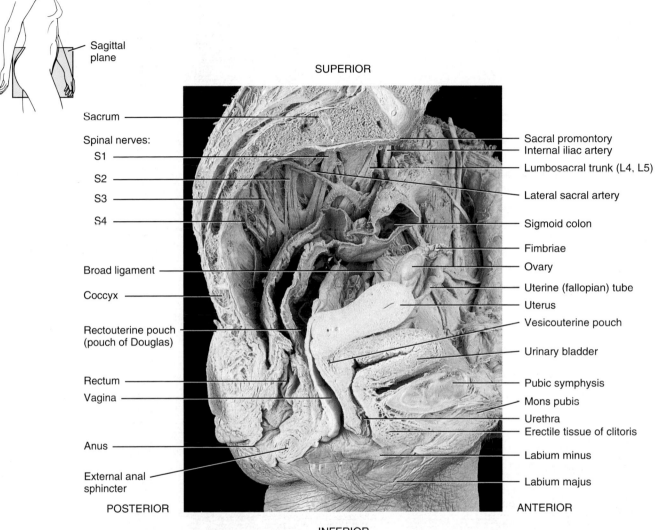

Sagittal plane

SUPERIOR

Sacrum

Spinal nerves:
S1
S2
S3
S4

Broad ligament

Coccyx

Rectouterine pouch
(pouch of Douglas)

Rectum

Vagina

Anus

External anal
sphincter

POSTERIOR

Sacral promontory
Internal iliac artery
Lumbosacral trunk (L4, L5)
Lateral sacral artery
Sigmoid colon
Fimbriae
Ovary
Uterine (fallopian) tube
Uterus
Vesicouterine pouch
Urinary bladder
Pubic symphysis
Mons pubis
Urethra
Erectile tissue of clitoris
Labium minus
Labium majus

ANTERIOR

INFERIOR

Sagittal section

FIGURE 14.8

Female reproductive organs

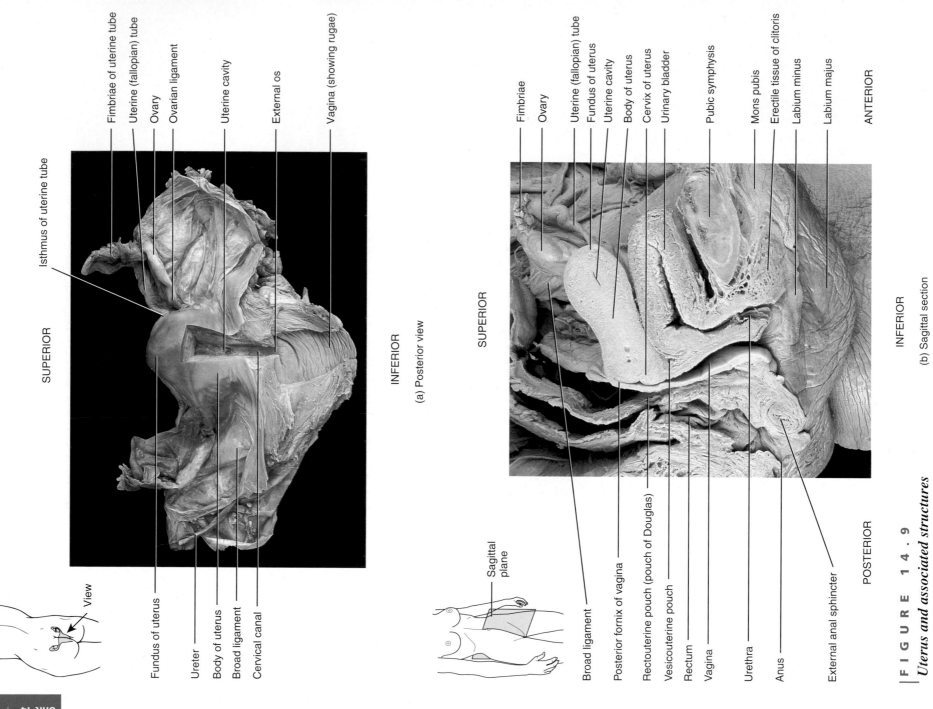

Fimbriae of uterine tube
Uterine (fallopian) tube
Ovary
Ovarian ligament
Uterine cavity
External os
Vagina (showing rugae)

Isthmus of uterine tube

SUPERIOR

INFERIOR

(a) Posterior view

View

Fundus of uterus
Ureter
Body of uterus
Broad ligament
Cervical canal

Fimbriae
Ovary
Uterine (fallopian) tube
Fundus of uterus
Uterine cavity
Body of uterus
Cervix of uterus
Urinary bladder
Pubic symphysis
Mons pubis
Erectile tissue of clitoris
Labium minus
Labium majus

ANTERIOR

SUPERIOR

INFERIOR

(b) Sagittal section

Sagittal plane

Broad ligament
Posterior fornix of vagina
Rectouterine pouch (pouch of Douglas)
Vesicouterine pouch
Rectum
Vagina
Urethra
Anus
External anal sphincter

POSTERIOR

FIGURE 14.9 *Uterus and associated structures*

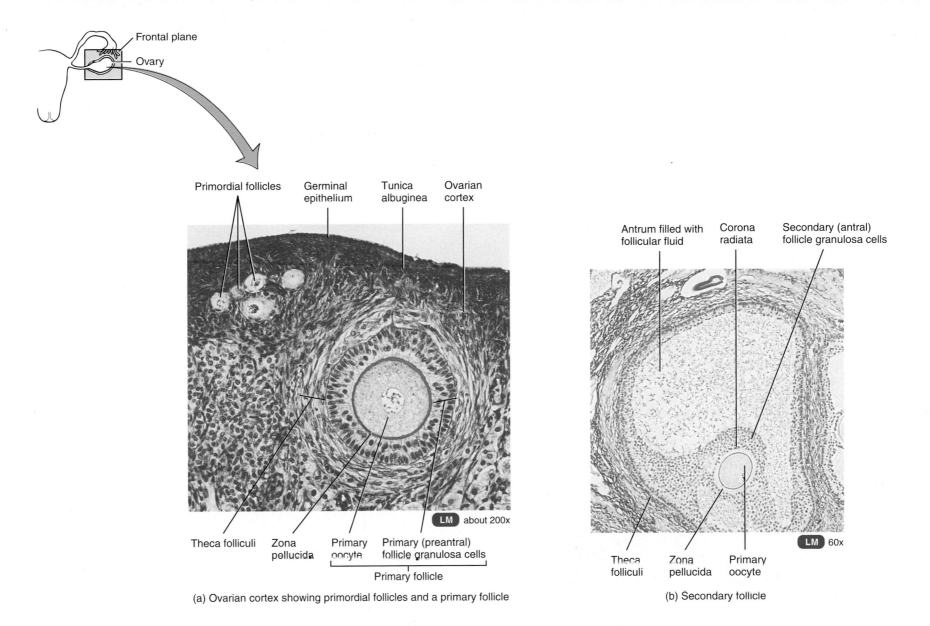

Frontal plane

Ovary

Primordial follicles Germinal epithelium Tunica albuginea Ovarian cortex

LM about 200x

Theca folliculi Zona pellucida Primary oocyte Primary (preantral) follicle granulosa cells

Primary follicle

(a) Ovarian cortex showing primordial follicles and a primary follicle

Antrum filled with follicular fluid Corona radiata Secondary (antral) follicle granulosa cells

LM 60x

Theca folliculi Zona pellucida Primary oocyte

(b) Secondary follicle

FIGURE 14.10

Histology of the ovary

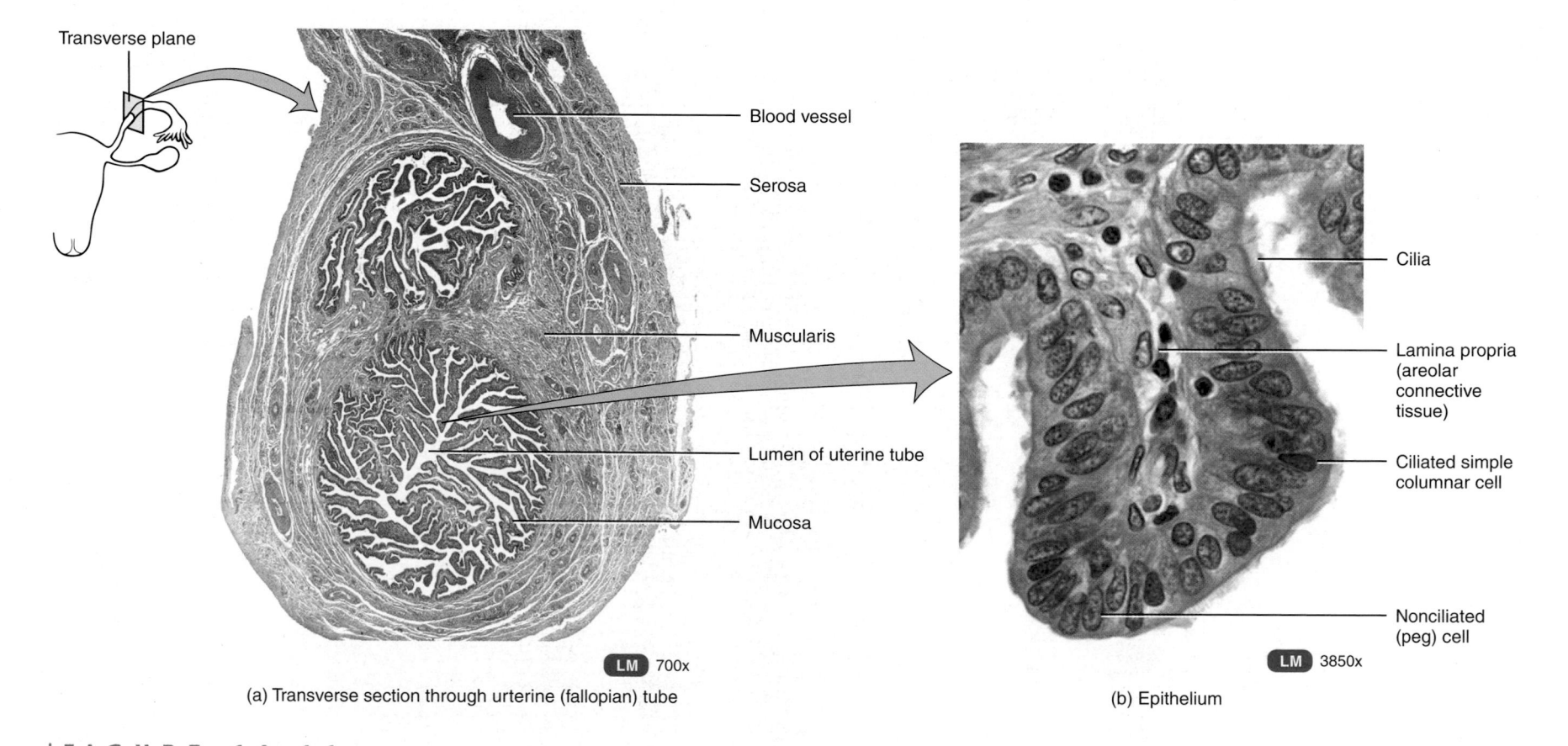

Transverse plane

Blood vessel

Serosa

Muscularis

Lumen of uterine tube

Mucosa

Cilia

Lamina propria
(areolar
connective
tissue)

Ciliated simple
columnar cell

Nonciliated
(peg) cell

LM 700x

LM 3850x

(a) Transverse section through urterine (fallopian) tube

(b) Epithelium

FIGURE 14.11

Histology of the uterine (fallopian) tube

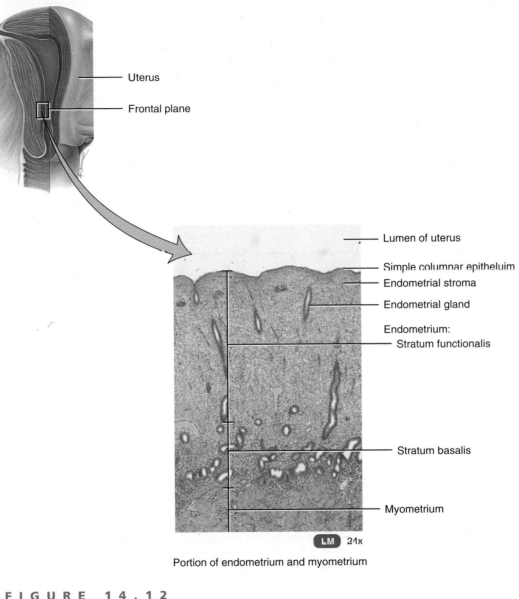

Uterus

Frontal plane

Lumen of uterus

Simple columnar epitheluim

Endometrial stroma

Endometrial gland

Endometrium:
Stratum functionalis

Stratum basalis

Myometrium

LM 24x

Portion of endometrium and myometrium

FIGURE 14.12
Histology of the uterus

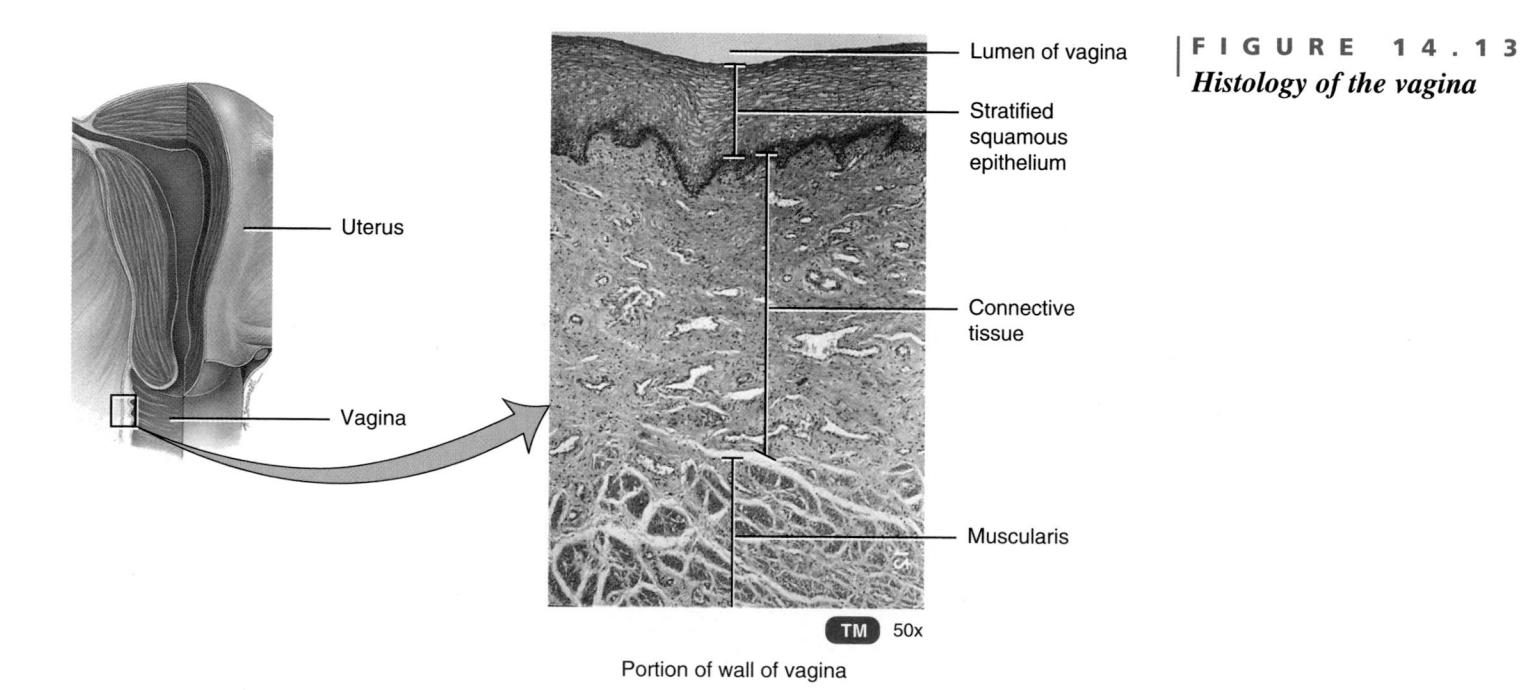

Uterus

Vagina

Lumen of vagina

Stratified
squamous
epithelium

Connective
tissue

Muscularis

TM 50x

Portion of wall of vagina

FIGURE 14.13
Histology of the vagina

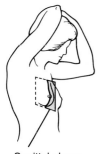
Sagittal plane

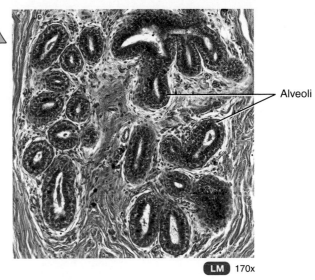

Alveoli

LM 170x

(b) Histology of section of a nonlactating
mammary gland

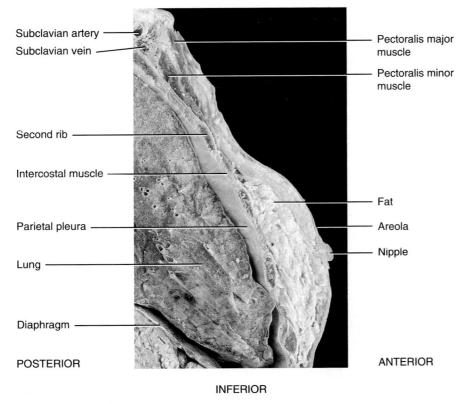

SUPERIOR

Subclavian artery

Subclavian vein

Second rib

Intercostal muscle

Parietal pleura

Lung

Diaphragm

POSTERIOR

Pectoralis major
muscle

Pectoralis minor
muscle

Fat

Areola

Nipple

ANTERIOR

INFERIOR

(a) Sagittal section of gross anatomy

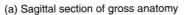

FIGURE 14.14
Mammary glands

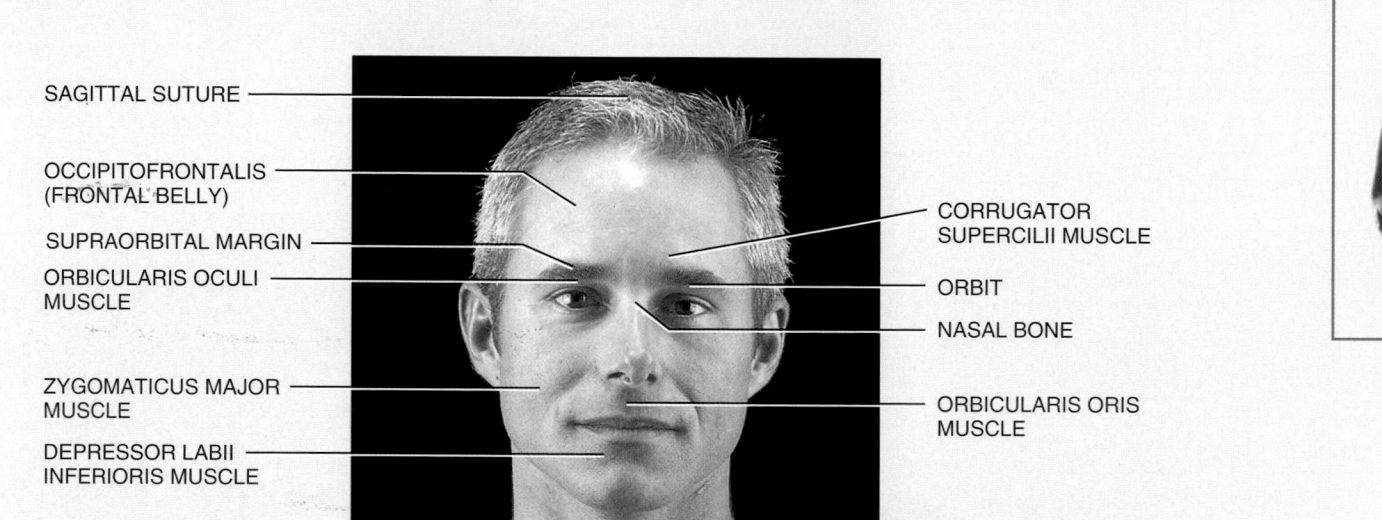

SAGITTAL SUTURE

OCCIPITOFRONTALIS
(FRONTAL BELLY)

SUPRAORBITAL MARGIN

ORBICULARIS OCULI
MUSCLE

ZYGOMATICUS MAJOR
MUSCLE

DEPRESSOR LABII
INFERIORIS MUSCLE

CORRUGATOR
SUPERCILII MUSCLE

ORBIT

NASAL BONE

ORBICULARIS ORIS
MUSCLE

(a) Anterior view of the head

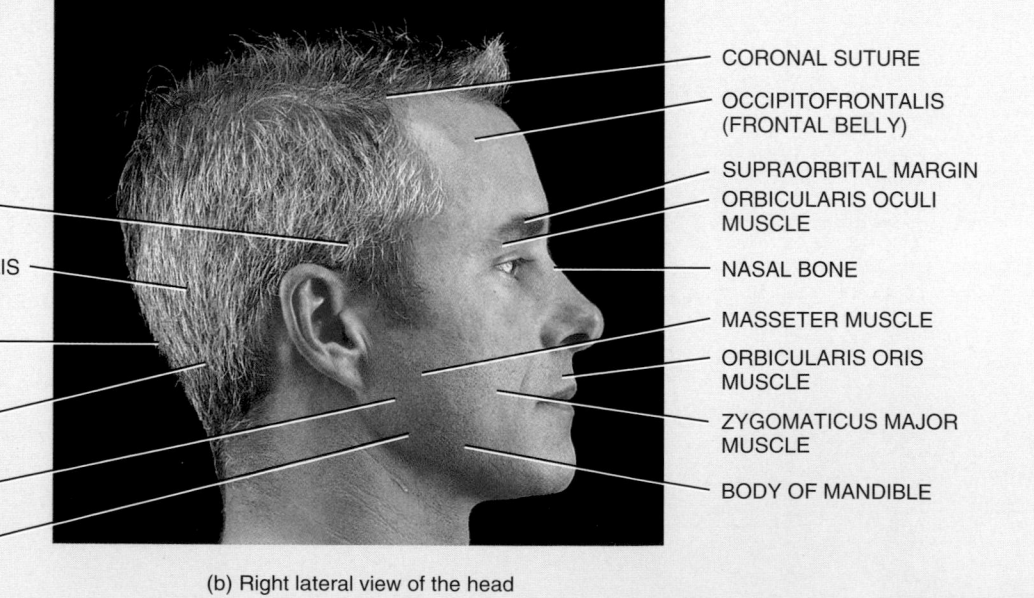

CORONAL SUTURE

OCCIPITOFRONTALIS
(FRONTAL BELLY)

SUPRAORBITAL MARGIN

ORBICULARIS OCULI
MUSCLE

NASAL BONE

MASSETER MUSCLE

ORBICULARIS ORIS
MUSCLE

ZYGOMATICUS MAJOR
MUSCLE

BODY OF MANDIBLE

TEMPORALIS
MUSCLE

OCCIPITOFRONTALIS
(OCCIPITAL BELLY)

EXTERNAL
OCCIPITAL
PROTUBERANCE

LAMBDOID
SUTURE

RAMUS OF
MANDIBLE

ANGLE OF
MANDIBLE

(b) Right lateral view of the head

FIGURE 15.1

Surface anatomy of the head and neck. Refer to Fig. 9.3a for the surface anatomy of the eye, Fig. 9.4a for the surface anatomy of the ear, and 9.1a for the surface anatomy of the nose.

FIGURE 15.1

Surface anatomy of the head and neck, continued

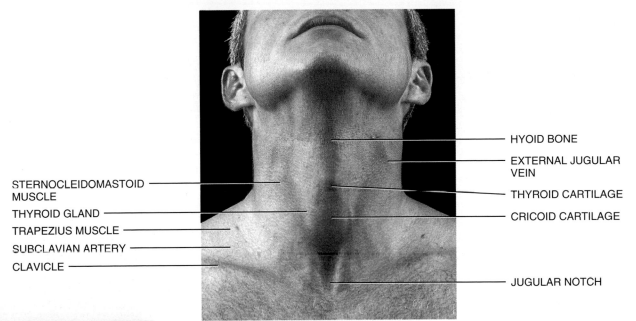

HYOID BONE

EXTERNAL JUGULAR VEIN

STERNOCLEIDOMASTOID MUSCLE

THYROID CARTILAGE

THYROID GLAND

CRICOID CARTILAGE

TRAPEZIUS MUSCLE

SUBCLAVIAN ARTERY

CLAVICLE

JUGULAR NOTCH

(c) Anterior view of the neck

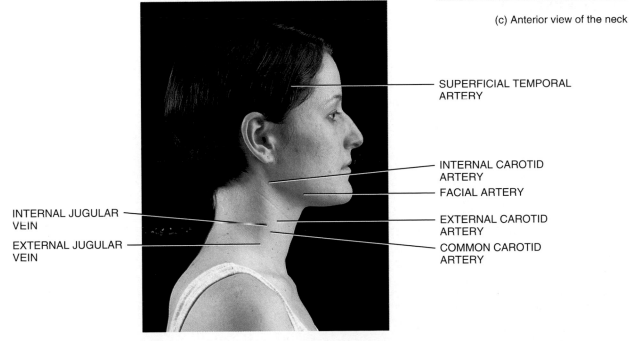

SUPERFICIAL TEMPORAL ARTERY

INTERNAL CAROTID ARTERY

FACIAL ARTERY

INTERNAL JUGULAR VEIN

EXTERNAL CAROTID ARTERY

EXTERNAL JUGULAR VEIN

COMMON CAROTID ARTERY

(d) Right lateral view of the neck

Unit 15 | 201

FIGURE 15.2 *Surface anatomy of the trunk*

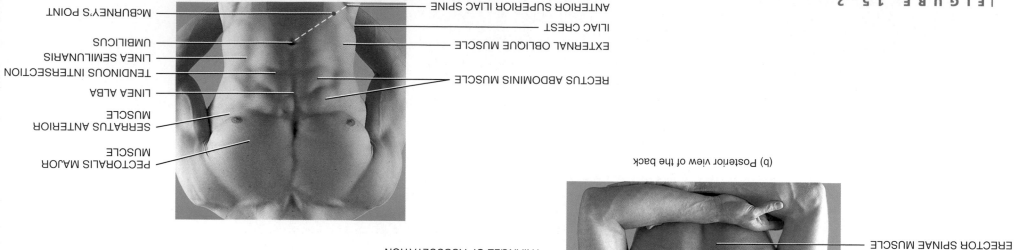

(c) Anterior view of abdomen

- McBURNEY'S POINT
- ANTERIOR SUPERIOR ILIAC SPINE
- ILIAC CREST
- UMBILICUS
- EXTERNAL OBLIQUE MUSCLE
- LINEA SEMILUNARIS
- TENDINOUS INTERSECTION
- RECTUS ABDOMINIS MUSCLE
- LINEA ALBA
- SERRATUS ANTERIOR MUSCLE
- PECTORALIS MAJOR MUSCLE

(b) Posterior view of the back

- ERECTOR SPINAE MUSCLE
- TRIANGLE OF AUSCULTATION
- POSTERIOR AXILLARY FOLD
- LATISSIMUS DORSI MUSCLE
- TERES MAJOR MUSCLE
- SCAPULA (VERTEBRAL BORDER)
- INFRASPINATUS MUSCLE
- VERTEBRAL SPINES
- TRAPEZIUS MUSCLE
- VERTEBRA PROMINENS

(a) Anterior view of the chest

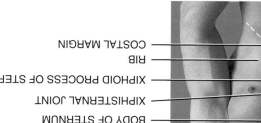
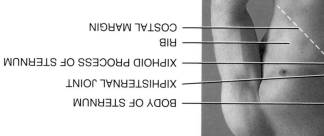

- COSTAL MARGIN
- RIB
- XIPHOID PROCESS OF STERNUM
- SERRATUS ANTERIOR MUSCLE
- NIPPLE
- XIPHISTERNAL JOINT
- ANTERIOR AXILLARY FOLD
- BODY OF STERNUM
- PECTORALIS MAJOR MUSCLE
- STERNAL ANGLE
- CLAVICLE
- MANUBRIUM OF STERNUM
- SUPRASTERNAL NOTCH

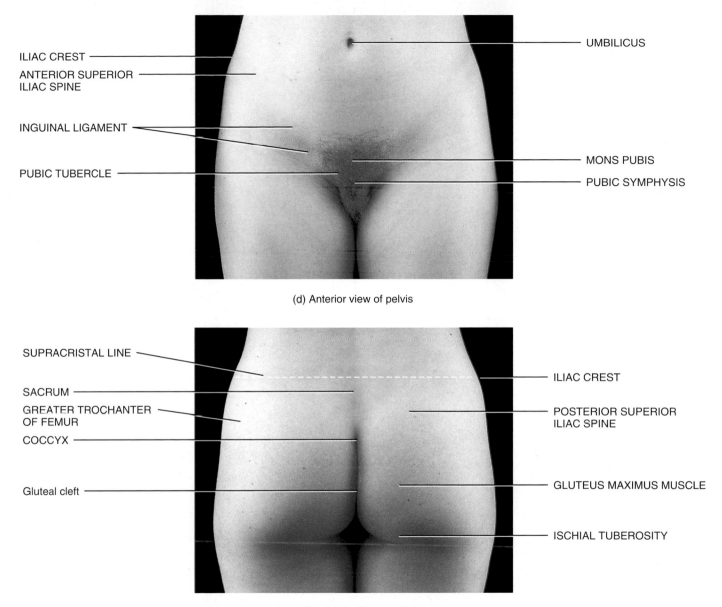

ILIAC CREST

ANTERIOR SUPERIOR
ILIAC SPINE

INGUINAL LIGAMENT

PUBIC TUBERCLE

UMBILICUS

MONS PUBIS

PUBIC SYMPHYSIS

(d) Anterior view of pelvis

SUPRACRISTAL LINE

SACRUM

GREATER TROCHANTER
OF FEMUR

COCCYX

Gluteal cleft

ILIAC CREST

POSTERIOR SUPERIOR
ILIAC SPINE

GLUTEUS MAXIMUS MUSCLE

ISCHIAL TUBEROSITY

(e) Posterior view of pelvis

FIGURE 15.2
Surface anatomy of the trunk, continued

ACROMIOCLAVICULAR
JOINT

ACROMION OF SCAPULA

CLAVICLE

SPINE OF SCAPULA

GREATER TUBERCLE
OF HUMERUS

DELTOID MUSCLE

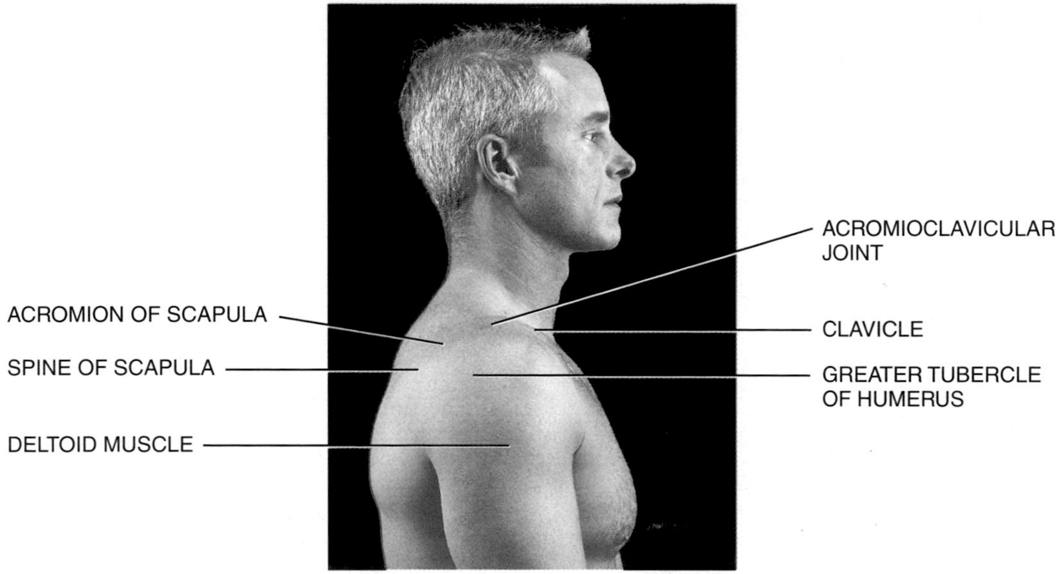

(a) Right lateral view of the shoulder

FIGURE 15.3
Surface anatomy of the upper limb

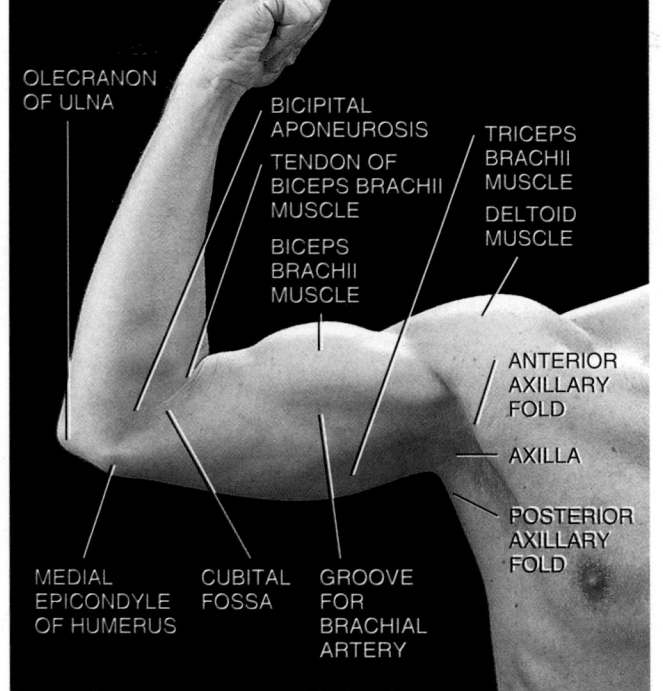

OLECRANON
OF ULNA

BICIPITAL
APONEUROSIS

TENDON OF
BICEPS BRACHII
MUSCLE

BICEPS
BRACHII
MUSCLE

TRICEPS
BRACHII
MUSCLE

DELTOID
MUSCLE

ANTERIOR
AXILLARY
FOLD

AXILLA

POSTERIOR
AXILLARY
FOLD

MEDIAL
EPICONDYLE
OF HUMERUS

CUBITAL
FOSSA

GROOVE
FOR
BRACHIAL
ARTERY

(b) Medial view of the arm and elbow

FIGURE 15.3
Surface anatomy of the upper limb, continued

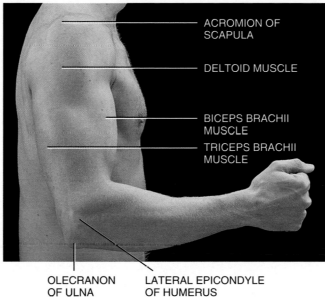

ACROMION OF SCAPULA

DELTOID MUSCLE

BICEPS BRACHII MUSCLE

TRICEPS BRACHII MUSCLE

OLECRANON OF ULNA

LATERAL EPICONDYLE OF HUMERUS

(c) Right lateral view of the arm and elbow

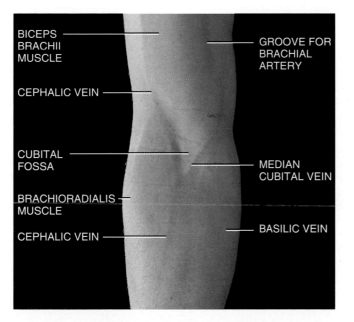

BICEPS BRACHII MUSCLE

CEPHALIC VEIN

CUBITAL FOSSA

BRACHIORADIALIS MUSCLE

CEPHALIC VEIN

GROOVE FOR BRACHIAL ARTERY

MEDIAN CUBITAL VEIN

BASILIC VEIN

(d) Anterior view of the cubital fossa

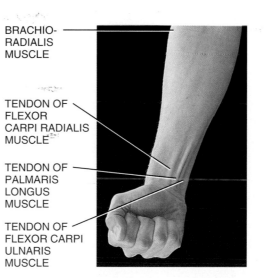

BRACHIORADIALIS MUSCLE

TENDON OF FLEXOR CARPI RADIALIS MUSCLE

TENDON OF PALMARIS LONGUS MUSCLE

TENDON OF FLEXOR CARPI ULNARIS MUSCLE

(e) Anterior aspect of the forearm and wrist

FIGURE 15.3

Surface anatomy of the upper limb, continued

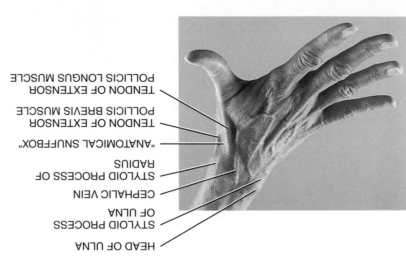

(g) Dorsum of the wrist

- TENDON OF EXTENSOR POLLICIS LONGUS MUSCLE
- TENDON OF EXTENSOR POLLICIS BREVIS MUSCLE
- "ANATOMICAL SNUFFBOX"
- STYLOID PROCESS OF RADIUS
- CEPHALIC VEIN
- STYLOID PROCESS OF ULNA
- HEAD OF ULNA

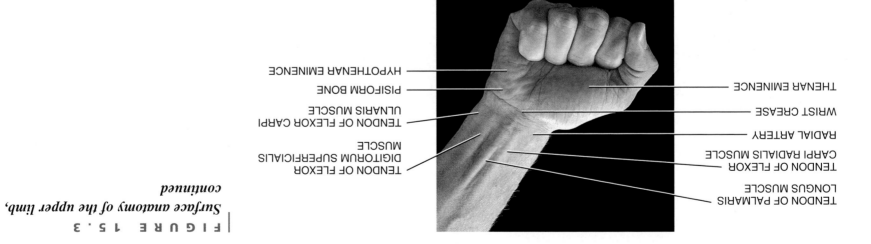

(f) Anterior aspect of the wrist

- HYPOTHENAR EMINENCE
- PISIFORM BONE
- TENDON OF FLEXOR CARPI ULNARIS MUSCLE
- TENDON OF FLEXOR DIGITORUM SUPERFICIALIS MUSCLE
- THENAR EMINENCE
- WRIST CREASE
- RADIAL ARTERY
- TENDON OF FLEXOR CARPI RADIALIS MUSCLE
- TENDON OF PALMARIS LONGUS MUSCLE

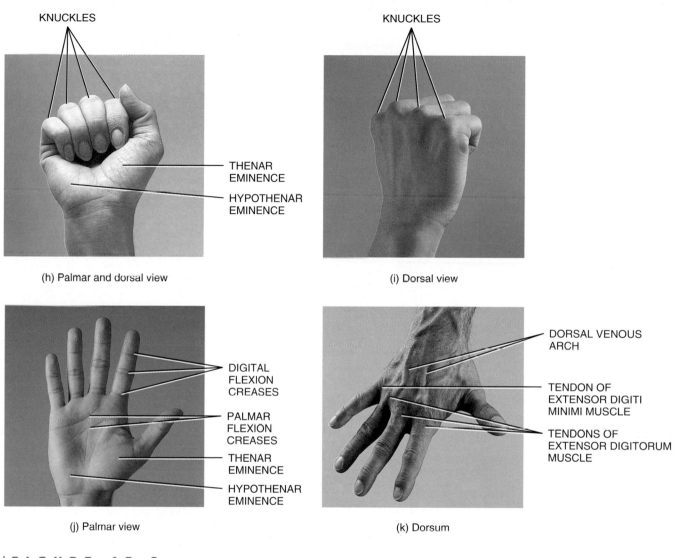

KNUCKLES

KNUCKLES

THENAR
EMINENCE

HYPOTHENAR
EMINENCE

(h) Palmar and dorsal view

(i) Dorsal view

DIGITAL
FLEXION
CREASES

PALMAR
FLEXION
CREASES

THENAR
EMINENCE

HYPOTHENAR
EMINENCE

DORSAL VENOUS
ARCH

TENDON OF
EXTENSOR DIGITI
MINIMI MUSCLE

TENDONS OF
EXTENSOR DIGITORUM
MUSCLE

(j) Palmar view

(k) Dorsum

FIGURE 15.3
Surface anatomy of the upper limb,
continued

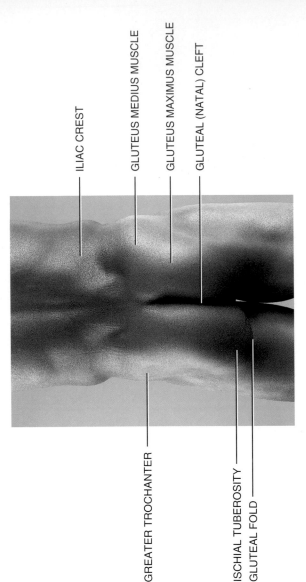

GREATER TROCHANTER

ISCHIAL TUBEROSITY
GLUTEAL FOLD

ILIAC CREST

GLUTEUS MEDIUS MUSCLE

GLUTEUS MAXIMUS MUSCLE

GLUTEAL (NATAL) CLEFT

(a) Posterior view of the buttocks and thigh

SEMITENDINOSUS AND
SEMIMEMBRANOSUS
MUSCLES

VASTUS LATERALIS
MUSCLE

BICEPS FEMORIS
MUSCLE

POPLITEAL FOSSA

TENDON OF
SEMITENDINOSUS
MUSCLE

GASTROCNEMIUS
MUSCLE (MEDIAL
AND LATERAL HEADS)

(c) Posterior view of the popliteal fossa

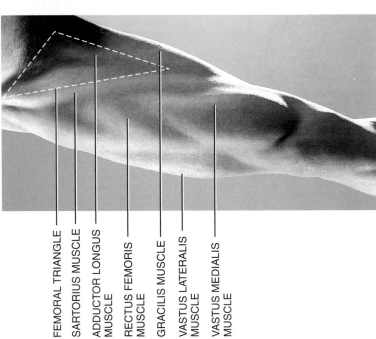

FEMORAL TRIANGLE

SARTORIUS MUSCLE

ADDUCTOR LONGUS
MUSCLE

RECTUS FEMORIS
MUSCLE

GRACILIS MUSCLE

VASTUS LATERALIS
MUSCLE

VASTUS MEDIALIS
MUSCLE

(b) Anterior view of the thigh

VASTUS LATERALIS MUSCLE

LATERAL CONDYLE OF FEMUR

PATELLA

LATERAL CONDYLE OF TIBIA

TIBIALIS ANTERIOR MUSCLE

VASTUS MEDIALIS MUSCLE

MEDIAL CONDYLE OF FEMUR

MEDIAL CONDYLE OF TIBIA

PATELLAR LIGAMENT

TIBIAL TUBEROSITY

(d) Anterior view of the knee

FIGURE 15.4

Surface anatomy of the lower limb

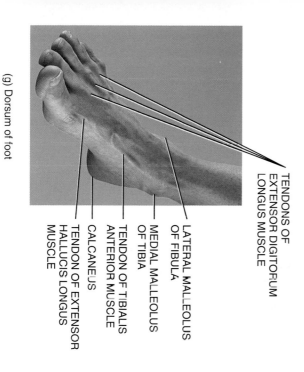

(g) Dorsum of foot

TENDONS OF
EXTENSOR DIGITORUM
LONGUS MUSCLE

LATERAL MALLEOLUS
OF FIBULA

MEDIAL MALLEOLUS
OF TIBIA

TENDON OF TIBIALIS
ANTERIOR MUSCLE

CALCANEUS

TENDON OF EXTENSOR
HALLUCIS LONGUS
MUSCLE

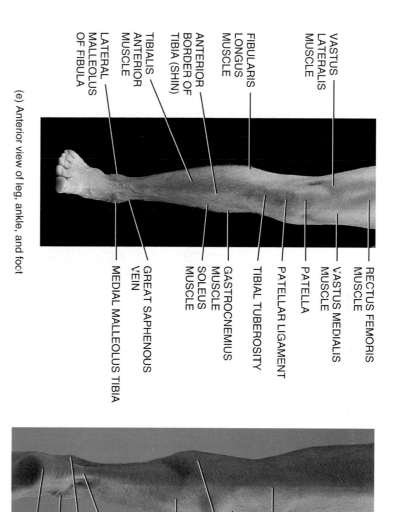

(e) Anterior view of leg, ankle, and foot

VASTUS
LATERALIS
MUSCLE

FIBULARIS
LONGUS
MUSCLE

ANTERIOR
BORDER OF
TIBIA (SHIN)

TIBIALIS
ANTERIOR
MUSCLE

LATERAL
MALLEOLUS
OF FIBULA

RECTUS FEMORIS
MUSCLE

VASTUS MEDIALIS
MUSCLE

PATELLA

PATELLAR LIGAMENT

TIBIAL TUBEROSITY

GASTROCNEMIUS
MUSCLE

SOLEUS
MUSCLE

GREAT SAPHENOUS
VEIN

MEDIAL MALLEOLUS TIBIA

(f) Posterior view of the leg and ankle

CALCANEUS

LATERAL MALLEOLUS
OF FIBULA

MEDIAL MALLEOLUS
OF TIBIA

CALCANEAL (ACHILLES)
TENDON

SOLEUS MUSCLE

GASTROCNEMIUS
MUSCLE (MEDIAL AND
LATERAL HEADS)

POPLITEAL FOSSA

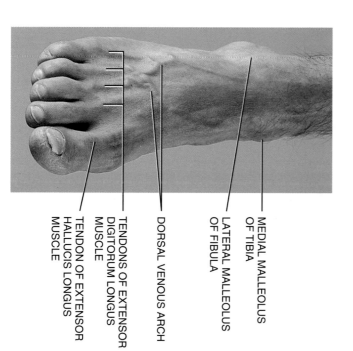

(h) Dorsum of foot

MEDIAL MALLEOLUS
OF TIBIA

LATERAL MALLEOLUS
OF FIBULA

DORSAL VENOUS ARCH

TENDONS OF EXTENSOR
DIGITORUM LONGUS
MUSCLE

TENDON OF EXTENSOR
HALLUCIS LONGUS
MUSCLE

F I G U R E   1 5 . 4
Surface anatomy of the lower limb,
continued

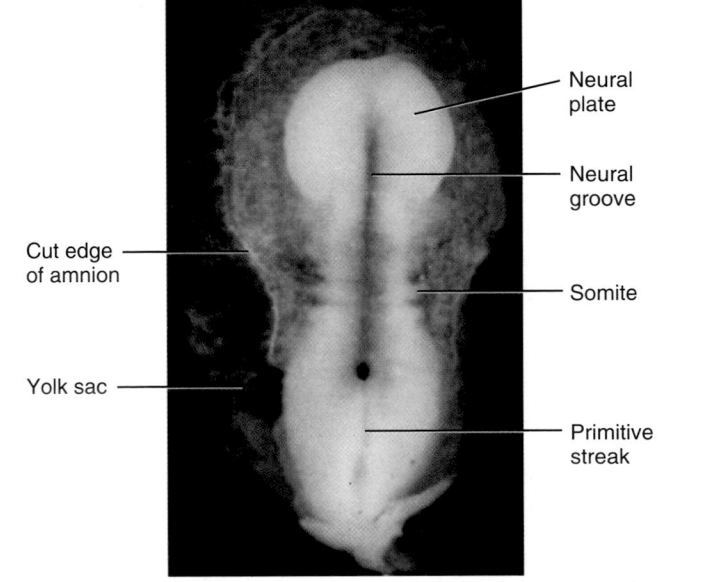

Neural
plate

Neural
groove

Cut edge
of amnion

Somite

Yolk sac

Primitive
streak

(a) 20-day embryo

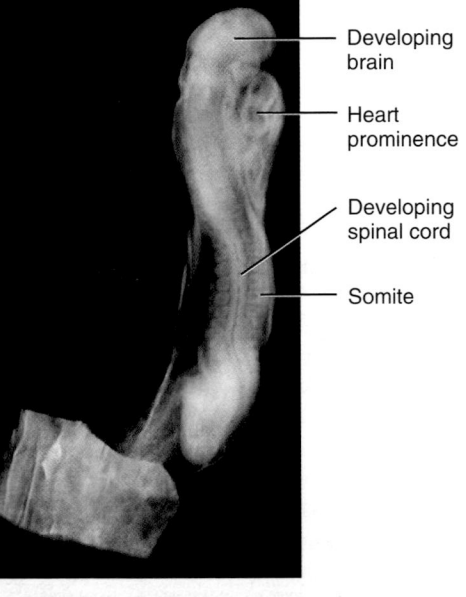

Developing
brain

Heart
prominence

Developing
spinal cord

Somite

(b) 24-day embryo

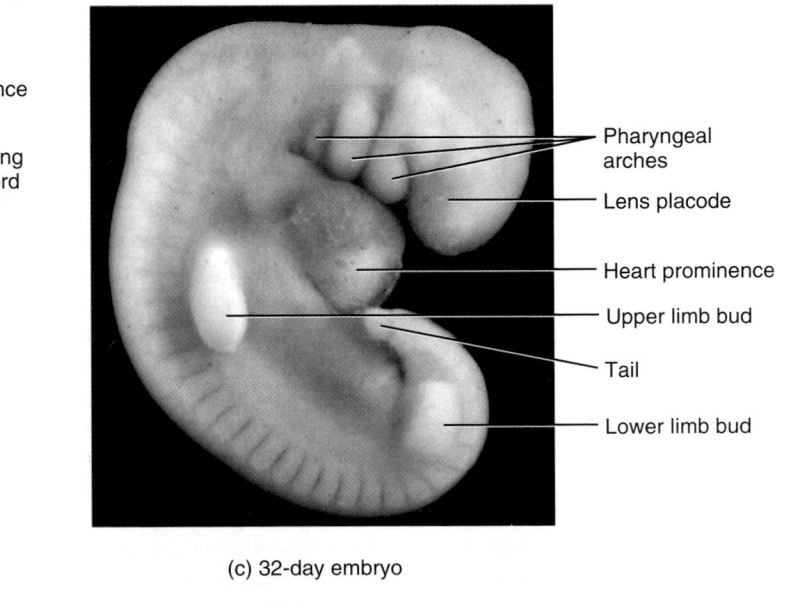

Pharyngeal
arches

Lens placode

Heart prominence

Upper limb bud

Tail

Lower limb bud

(c) 32-day embryo

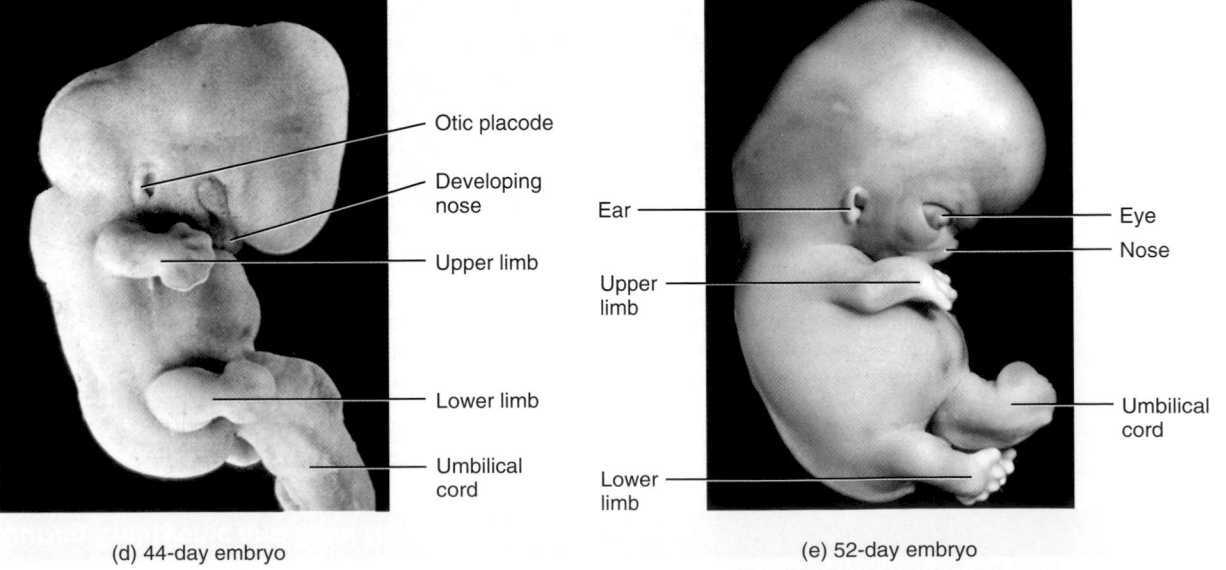

Otic placode

Developing
nose

Upper limb

Lower limb

Umbilical
cord

(d) 44-day embryo

Ear

Upper
limb

Lower
limb

Eye

Nose

Umbilical
cord

(e) 52-day embryo

FIGURE 16.1

Representative stages of embryonic and fetal development (not shown at their actual sizes)

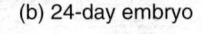

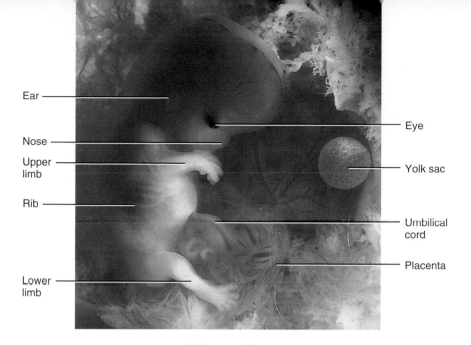

Ear

Nose

Upper
limb

Rib

Lower
limb

Eye

Yolk sac

Umbilical
cord

Placenta

(f) Ten-week fetus

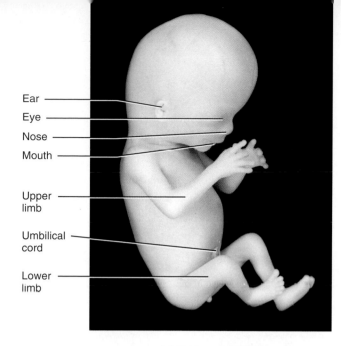

Ear

Eye

Nose

Mouth

Upper
limb

Umbilical
cord

Lower
limb

(g) Thirteen-week fetus

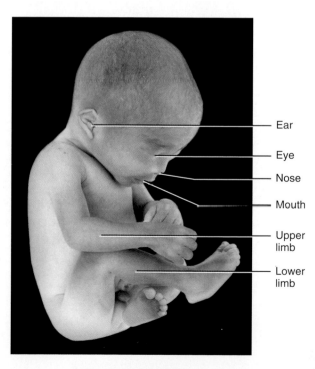

Ear

Eye

Nose

Mouth

Upper
limb

Lower
limb

(h) Twenty-six-week fetus

FIGURE 16.1
(continued)

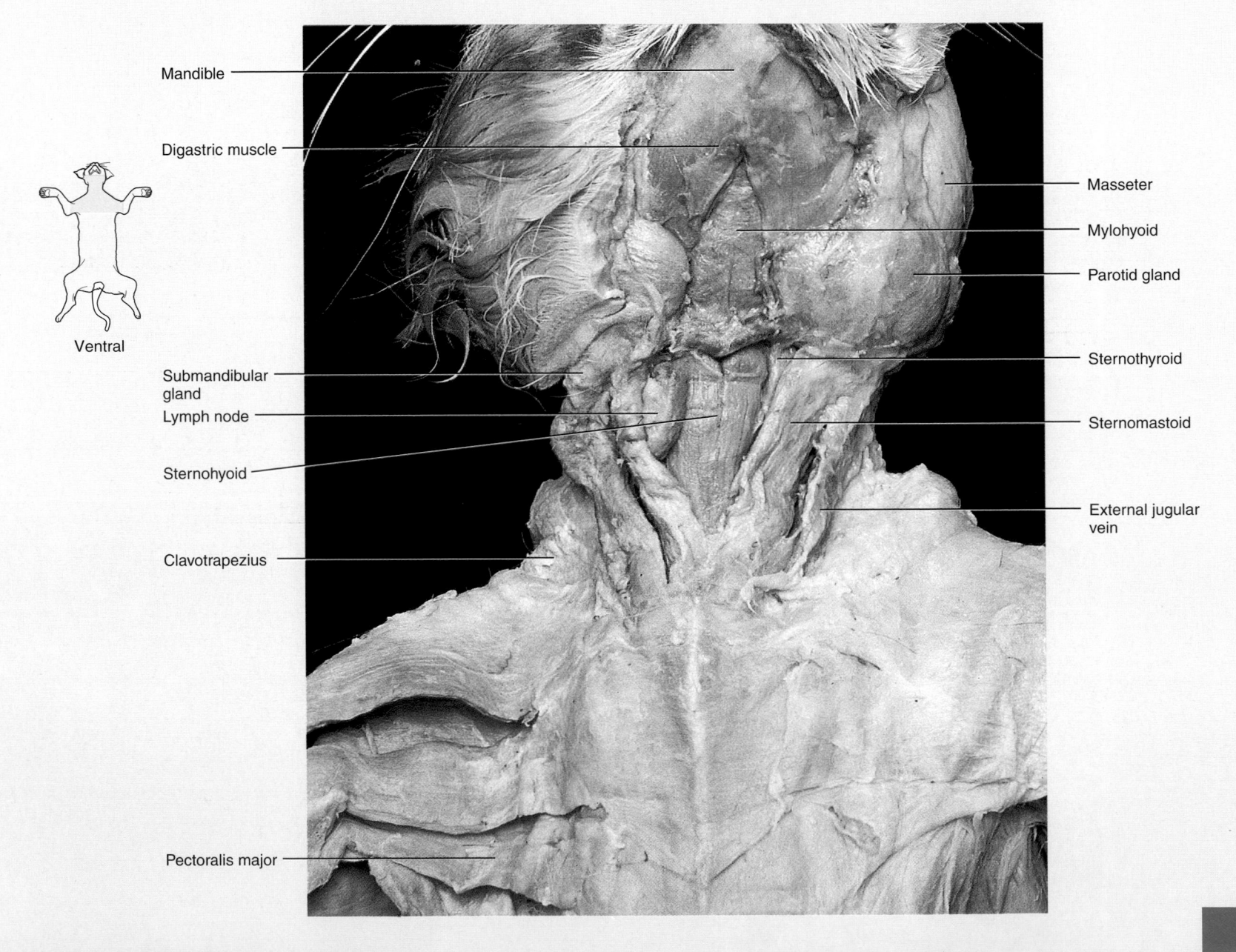

Mandible

Digastric muscle

Masseter

Mylohyoid

Parotid gland

Ventral

Sternothyroid

Submandibular gland

Lymph node

Sternomastoid

Sternohyoid

External jugular vein

Clavotrapezius

Pectoralis major

214

FIGURE A.1

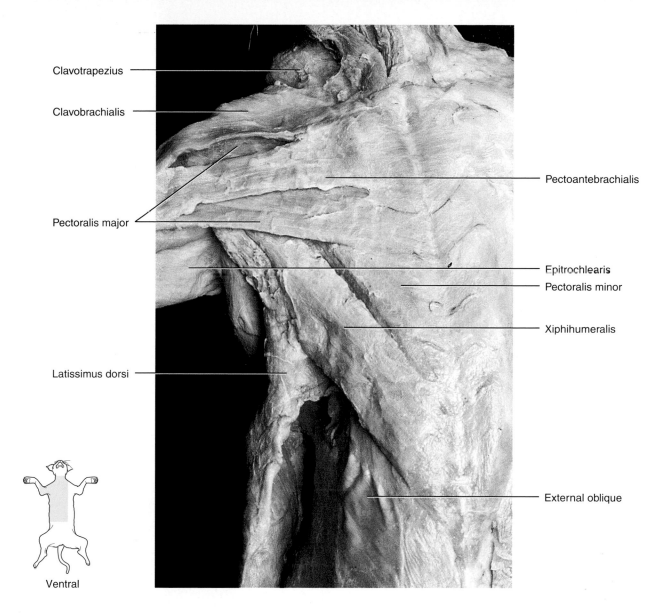

Clavotrapezius

Clavobrachialis

Pectoralis major

Latissimus dorsi

Pectoantebrachialis

Epitrochlearis

Pectoralis minor

Xiphihumeralis

External oblique

Ventral

FIGURE A.2A
Superficial muscles of the chest: ventral view

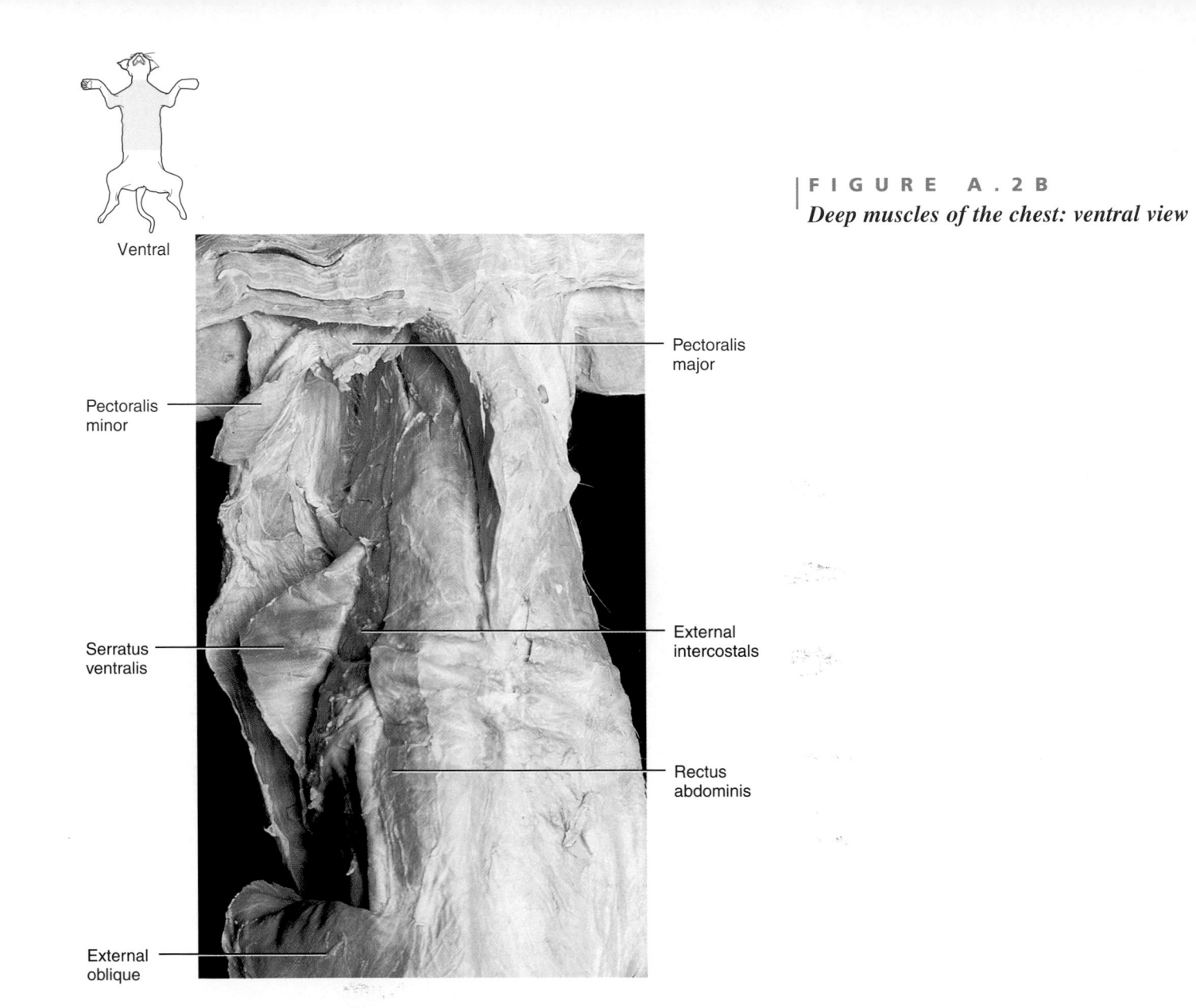

Ventral

Pectoralis
minor

Serratus
ventralis

External
oblique

Pectoralis
major

External
intercostals

Rectus
abdominis

Deep muscles of the chest: ventral view

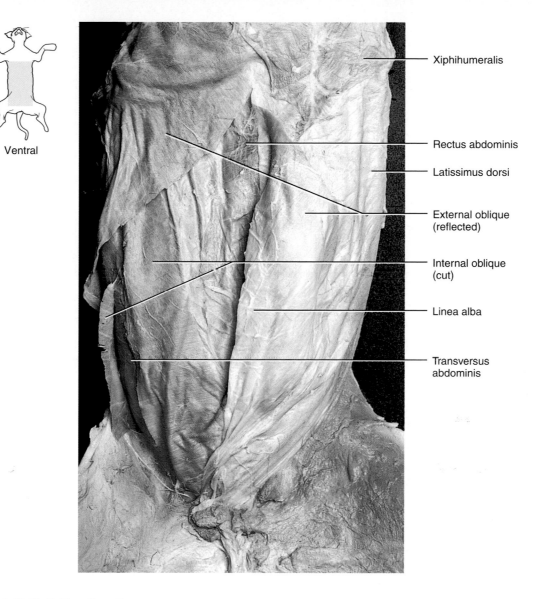

Ventral

Xiphihumeralis

Rectus abdominis

Latissimus dorsi

External oblique
(reflected)

Internal oblique
(cut)

Linea alba

Transversus
abdominis

FIGURE A.3
Muscles of the abdomen: ventral view

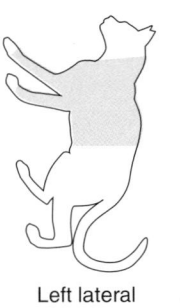

FIGURE A.4A

Superficial muscles of the shoulder: left lateral view

Left lateral

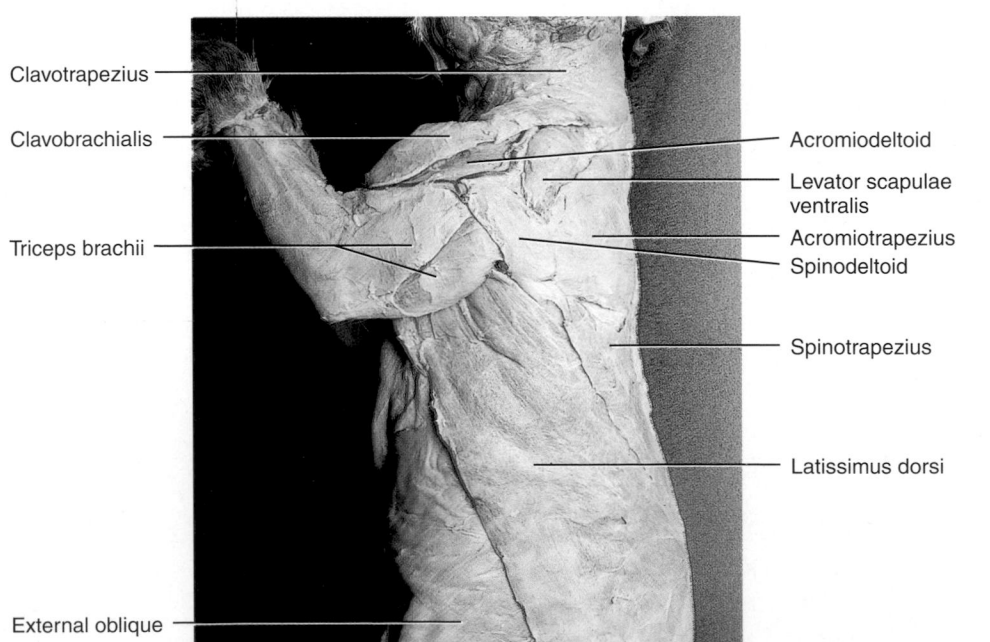

Clavotrapezius

Clavobrachialis

Triceps brachii

External oblique

Acromiodeltoid

Levator scapulae ventralis

Acromiotrapezius

Spinodeltoid

Spinotrapezius

Latissimus dorsi

Left lateral

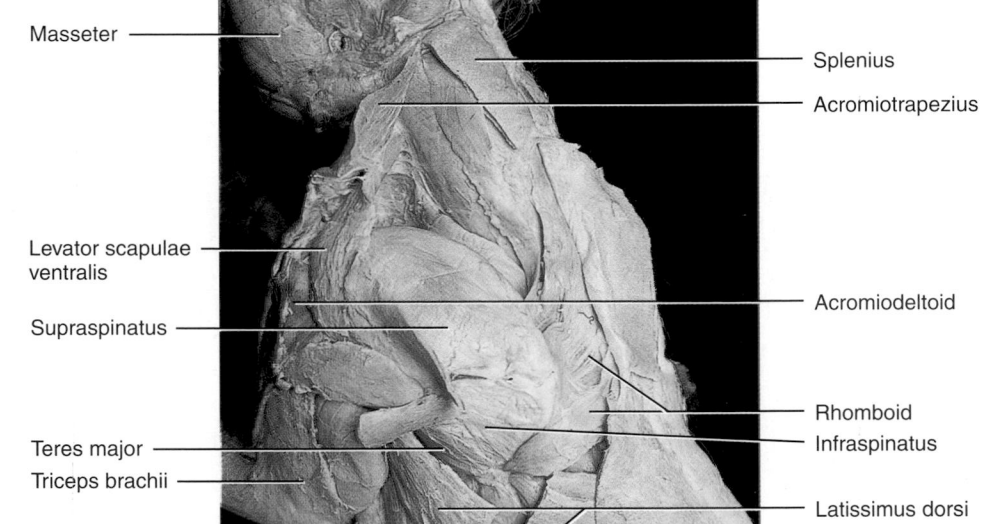

Masseter

Levator scapulae ventralis

Supraspinatus

Teres major

Triceps brachii

Splenius

Acromiotrapezius

Acromiodeltoid

Rhomboid

Infraspinatus

Latissimus dorsi

FIGURE A.4B

Deep muscles of the shoulder: left lateral view

Dorsal

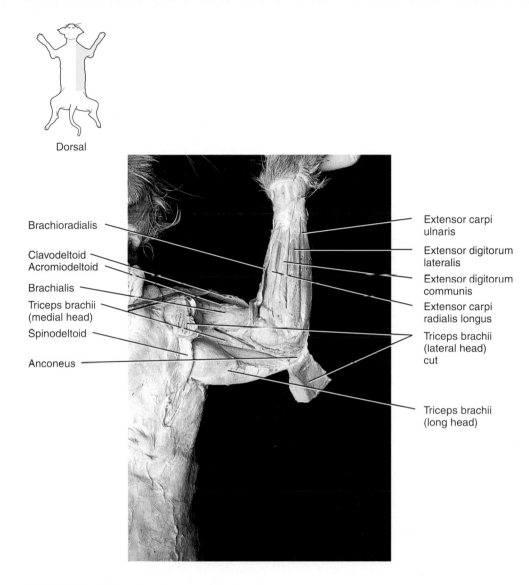

Brachioradialis

Clavodeltoid
Acromiodeltoid

Brachialis

Triceps brachii
(medial head)

Spinodeltoid

Anconeus

Extensor carpi
ulnaris

Extensor digitorum
lateralis

Extensor digitorum
communis

Extensor carpi
radialis longus

Triceps brachii
(lateral head)
cut

Triceps brachii
(long head)

FIGURE A.5
Muscles of the arm and forearm: lateral view

Muscles of the arm and forearm: medial view

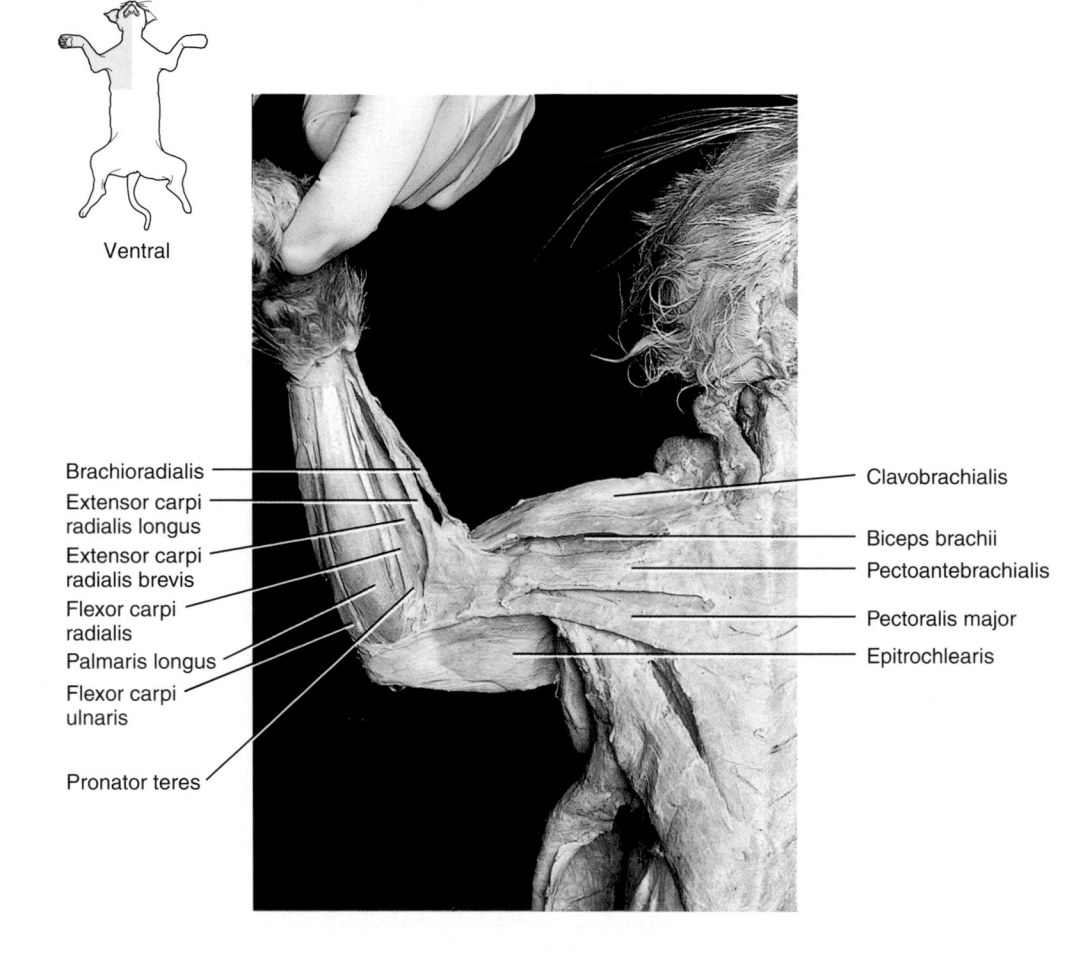

Ventral

Brachioradialis

Extensor carpi
radialis longus

Extensor carpi
radialis brevis

Flexor carpi
radialis

Palmaris longus

Flexor carpi
ulnaris

Pronator teres

Clavobrachialis

Biceps brachii

Pectoantebrachialis

Pectoralis major

Epitrochlearis

Right lateral

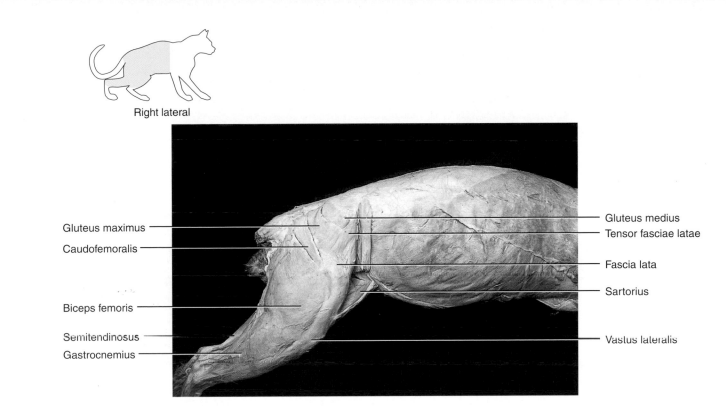

Gluteus maximus

Caudofemoralis

Biceps femoris

Semitendinosus

Gastrocnemius

Gluteus medius

Tensor fasciae latae

Fascia lata

Sartorius

Vastus lateralis

F I G U R E   A . 7

Superficial muscles of the thigh: right lateral view

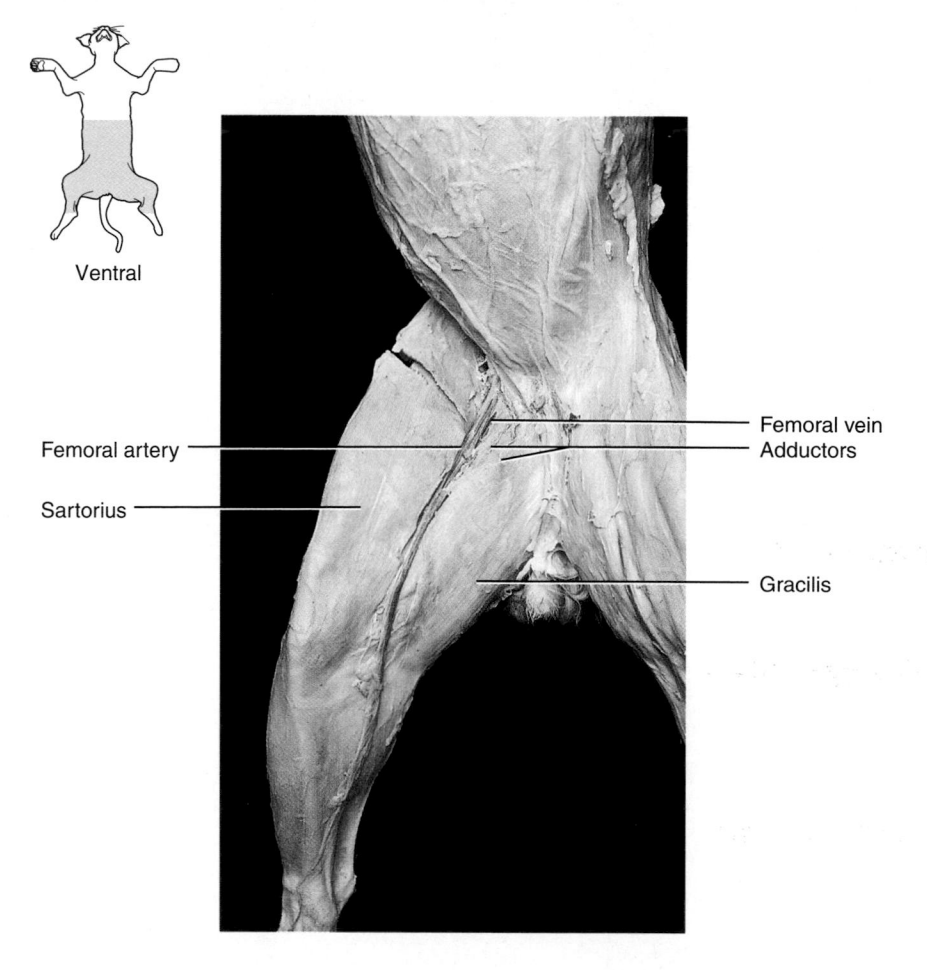

Ventral

Femoral artery

Sartorius

Femoral vein
Adductors

Gracilis

FIGURE A.8A
Superficial muscles of the thigh: medial view

FIGURE A.8B
Deep muscles of the thigh: medial view

Ventral

Sartorius
Iliopsoas
Femoral artery
Vastus lateralis
Rectus femoris
Sartorius

Pectineus
Adductor longus
Adductor femoris
Gracilis (cut)
Vastus medialis
Semimembranosus
Gastocnemius

Right lateral

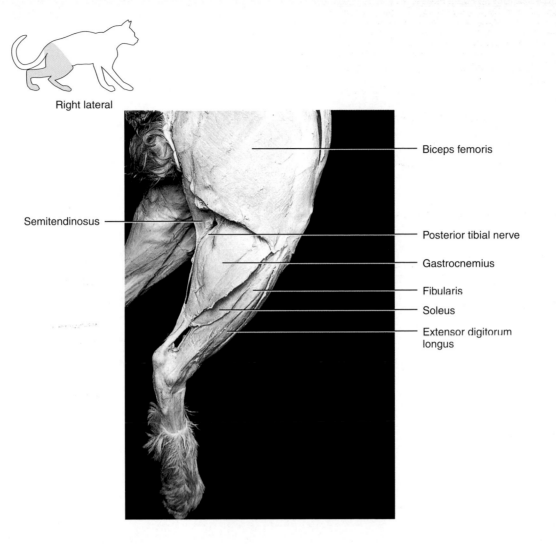

Biceps femoris

Semitendinosus

Posterior tibial nerve

Gastrocnemius

Fibularis

Soleus

Extensor digitorum
longus

FIGURE A.9
Superficial muscles of the leg; right lateral view

Superficial muscles of the leg: medial view

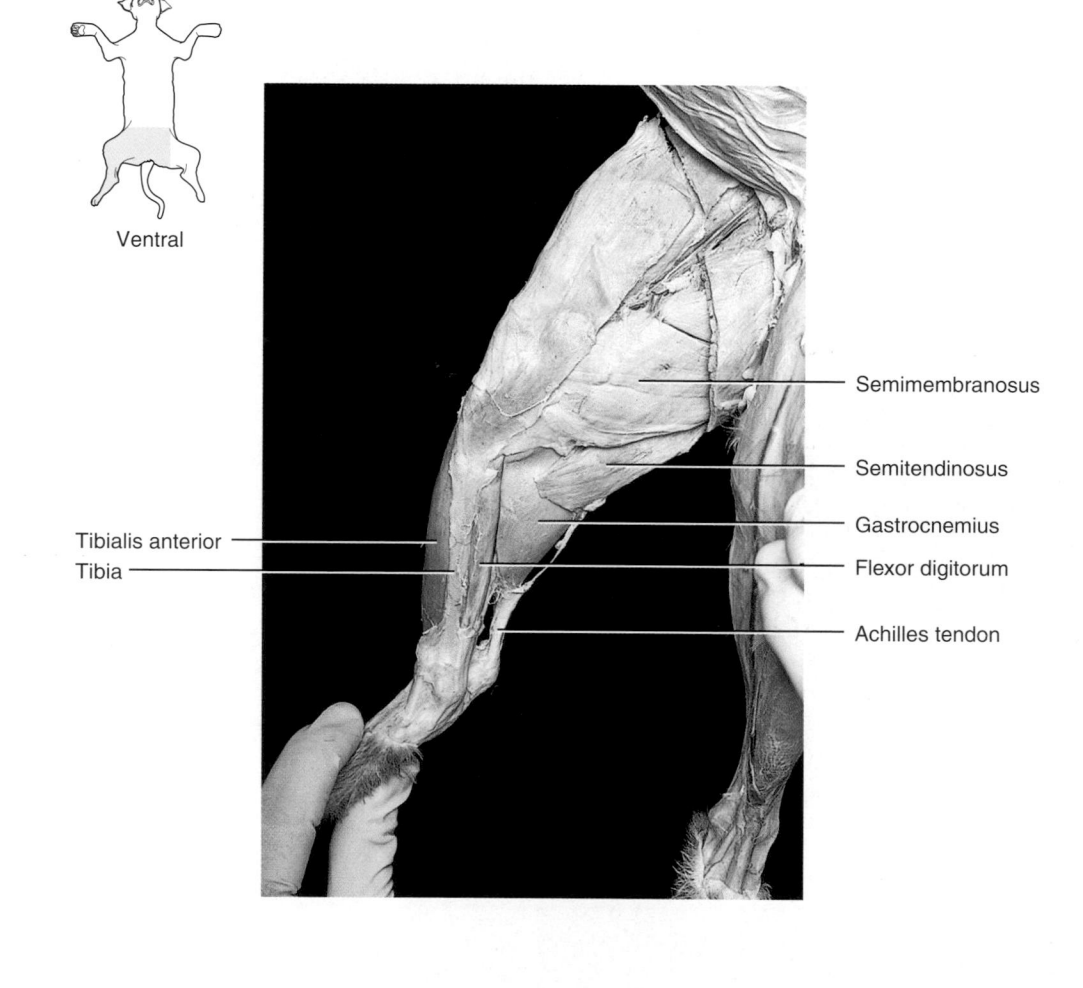

Ventral

Semimembranosus

Semitendinosus

Tibialis anterior

Gastrocnemius

Tibia

Flexor digitorum

Achilles tendon

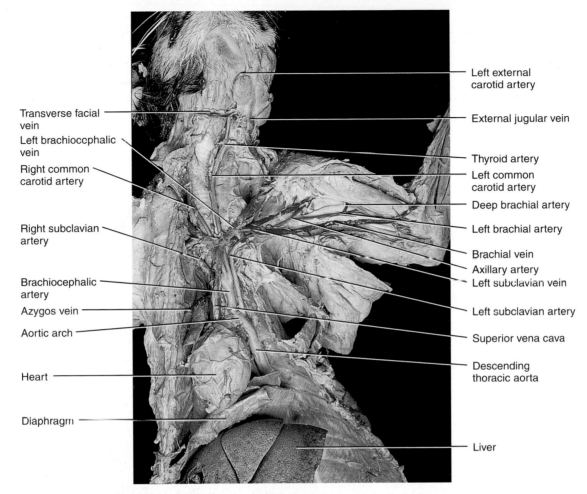

Transverse facial vein

Left brachiocephalic vein

Right common carotid artery

Right subclavian artery

Brachiocephalic artery

Azygos vein

Aortic arch

Heart

Diaphragm

Left external carotid artery

External jugular vein

Thyroid artery

Left common carotid artery

Deep brachial artery

Left brachial artery

Brachial vein

Axillary artery

Left subclavian vein

Left subclavian artery

Superior vena cava

Descending thoracic aorta

Liver

Ventral view

FIGURE A.11
Blood vessels above the diaphragm

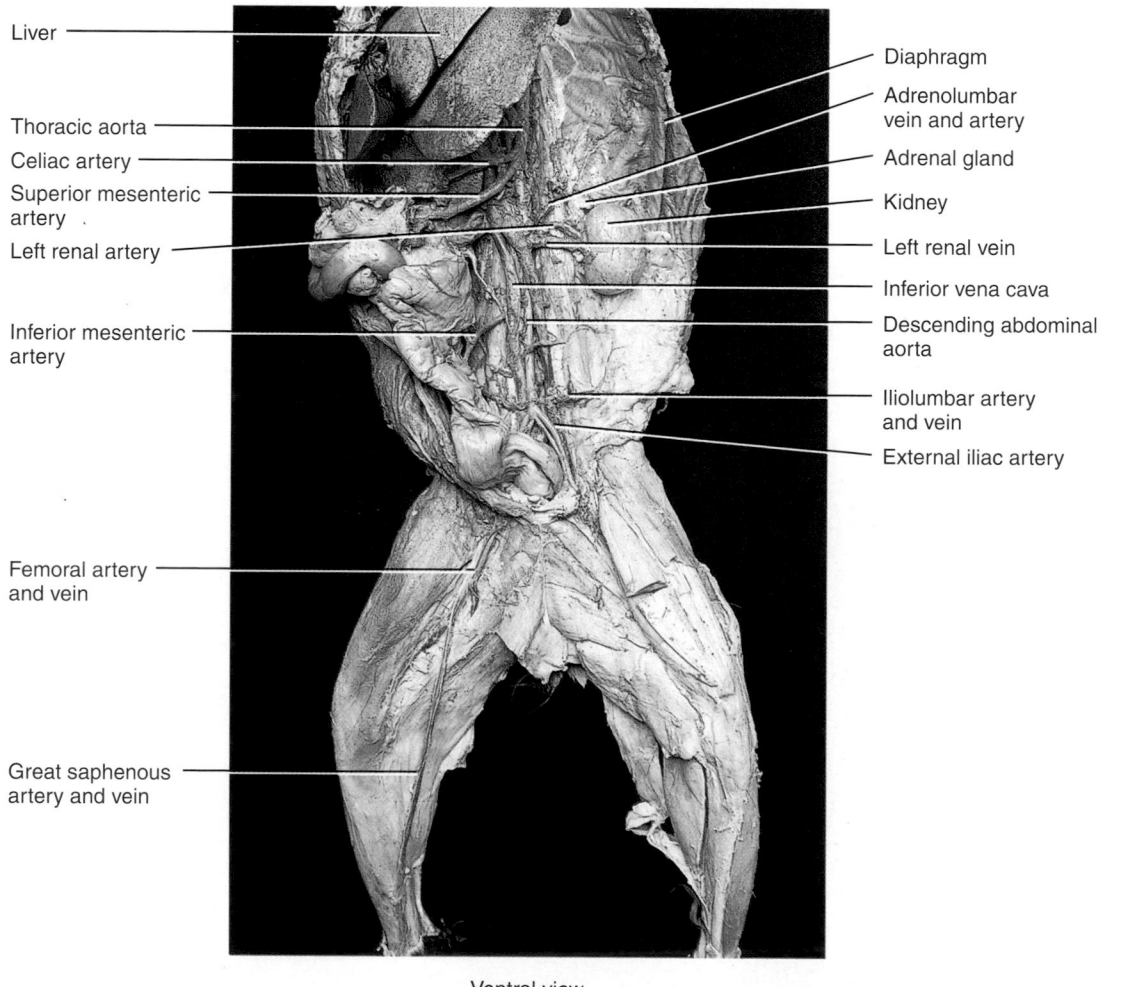

Liver

Thoracic aorta

Celiac artery

Superior mesenteric artery

Left renal artery

Inferior mesenteric artery

Femoral artery and vein

Great saphenous artery and vein

Diaphragm

Adrenolumbar vein and artery

Adrenal gland

Kidney

Left renal vein

Inferior vena cava

Descending abdominal aorta

Iliolumbar artery and vein

External iliac artery

Ventral view

FIGURE A.12
Blood vessels below the diaphragm

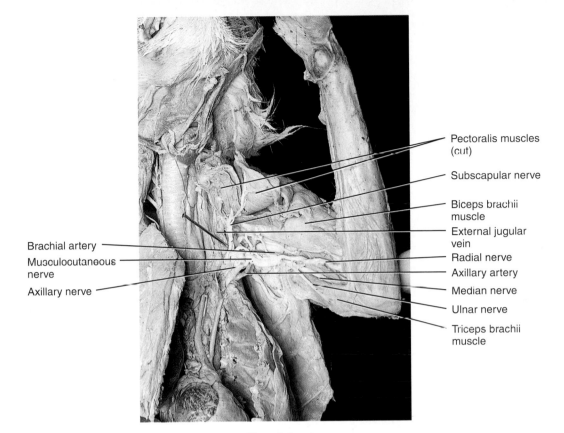

Pectoralis muscles (cut)

Subscapular nerve

Biceps brachii muscle

External jugular vein

Brachial artery

Musculocutaneous nerve

Axillary nerve

Radial nerve

Axillary artery

Median nerve

Ulnar nerve

Triceps brachii muscle

Ventral view

| FIGURE A . 1 3
Brachial plexus and associated blood vessels

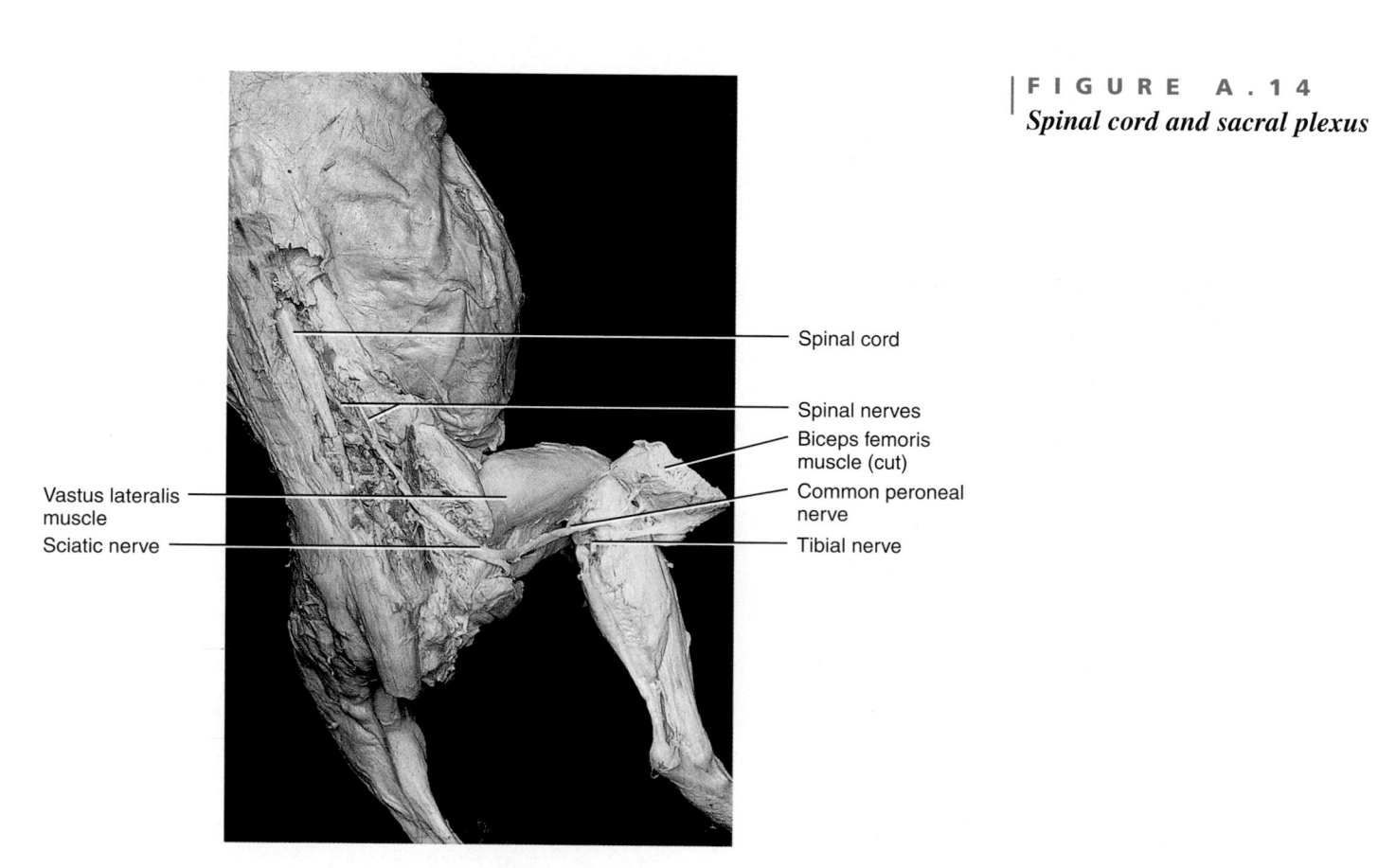

FIGURE A.14
Spinal cord and sacral plexus

Spinal cord

Spinal nerves

Biceps femoris
muscle (cut)

Common peroneal
nerve

Tibial nerve

Vastus lateralis
muscle

Sciatic nerve

Dorsal view

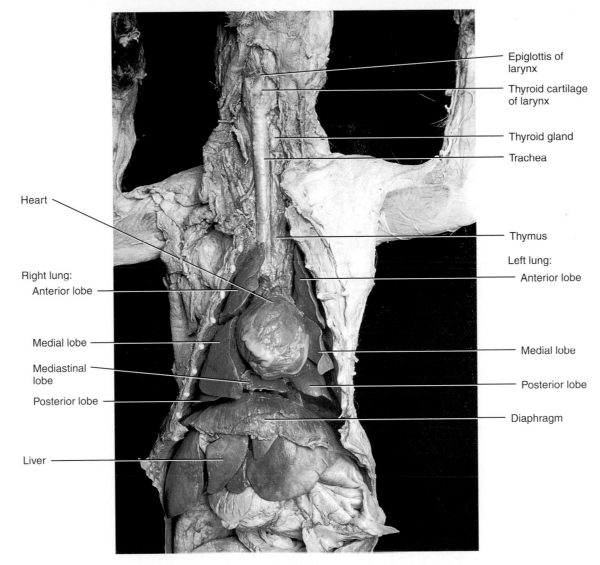

Epiglottis of
larynx

Thyroid cartilage
of larynx

Thyroid gland

Trachea

Heart

Thymus

Left lung:

Right lung:

Anterior lobe

Anterior lobe

Medial lobe

Medial lobe

Mediastinal
lobe

Posterior lobe

Posterior lobe

Diaphragm

Liver

Ventral view

FIGURE A.15
Respiratory system

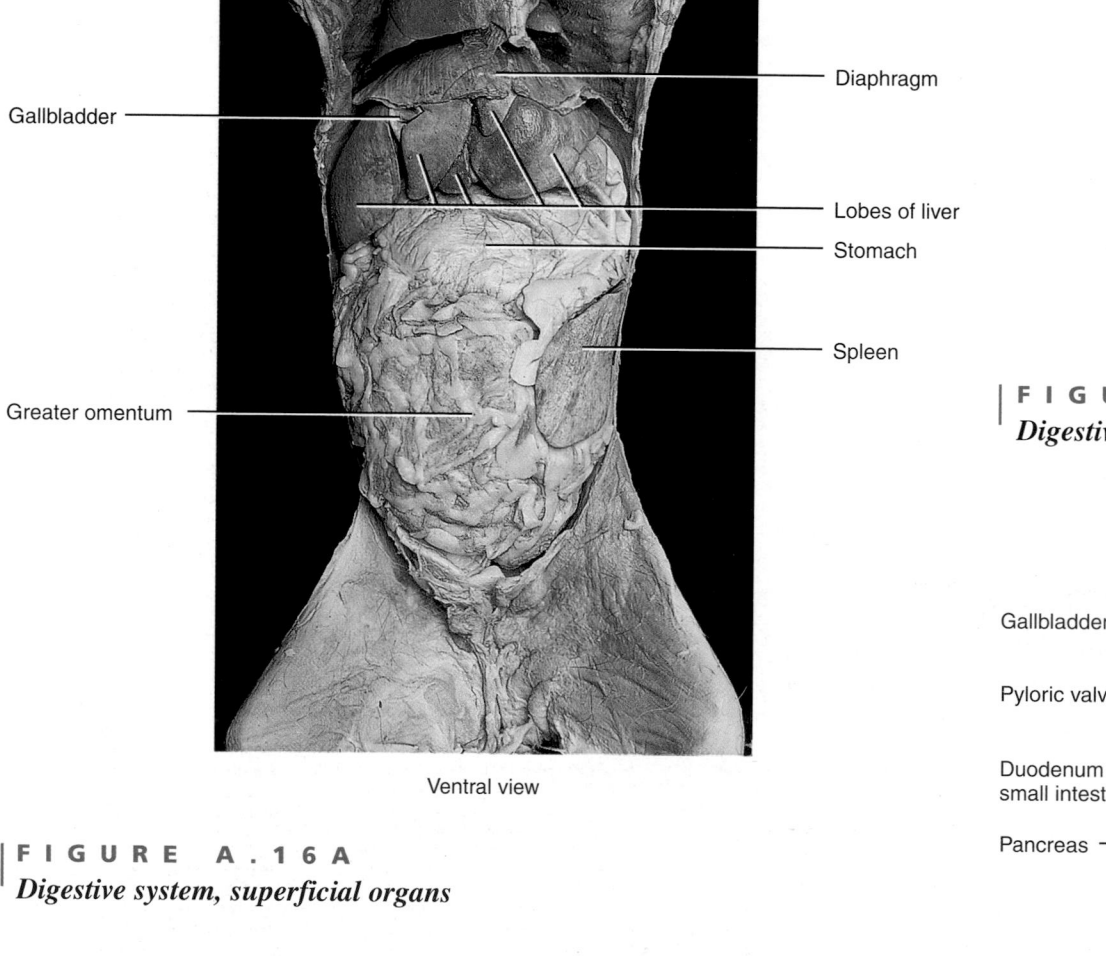

Gallbladder

Diaphragm

Lobes of liver

Stomach

Spleen

Greater omentum

Ventral view

FIGURE A.16A
Digestive system, superficial organs

FIGURE A.16B
Digestive system, deep organs

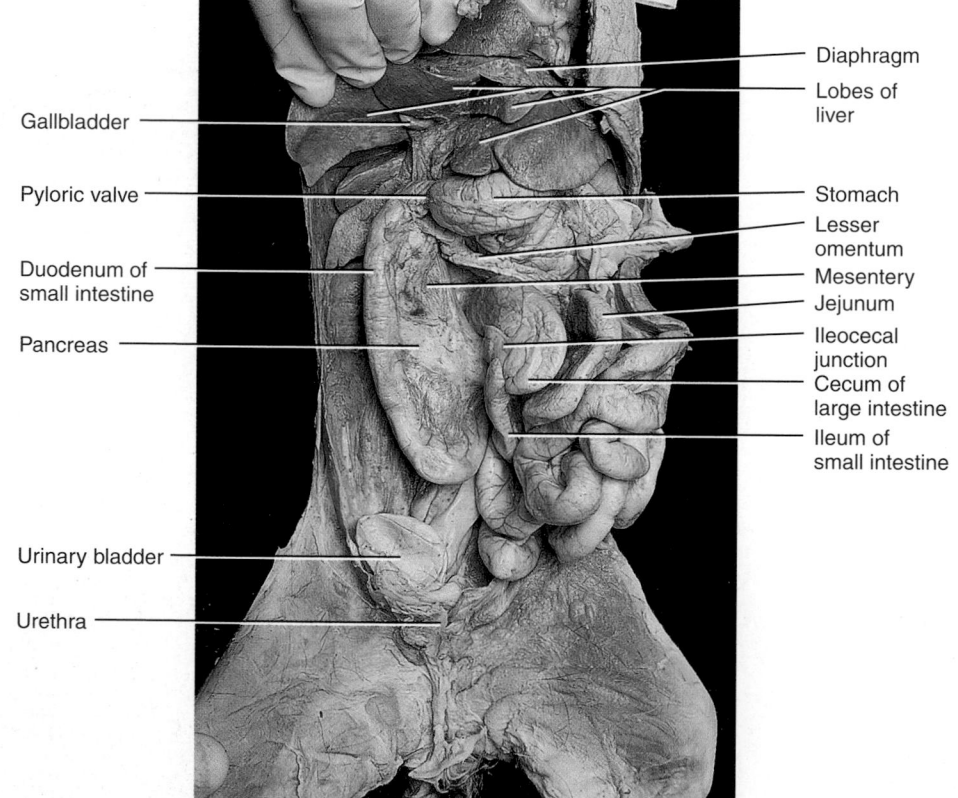

Gallbladder

Pyloric valve

Duodenum of
small intestine

Pancreas

Urinary bladder

Urethra

Diaphragm

Lobes of
liver

Stomach

Lesser
omentum

Mesentery

Jejunum

Ileocecal
junction

Cecum of
large intestine

Ileum of
small intestine

Ventral view

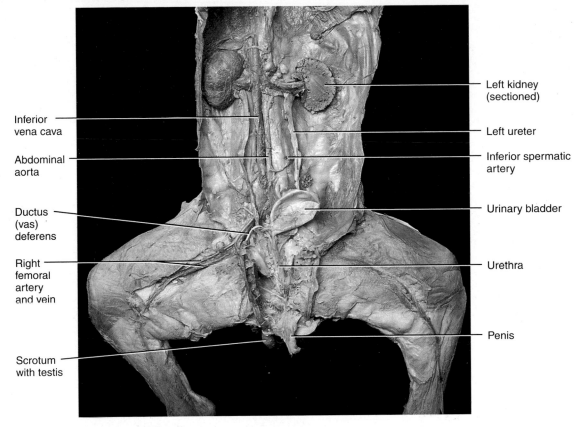

Inferior
vena cava

Abdominal
aorta

Ductus
(vas)
deferens

Right
femoral
artery
and vein

Scrotum
with testis

Left kidney
(sectioned)

Left ureter

Inferior spermatic
artery

Urinary bladder

Urethra

Penis

Ventral view

FIGURE A.17A
Male urinary system

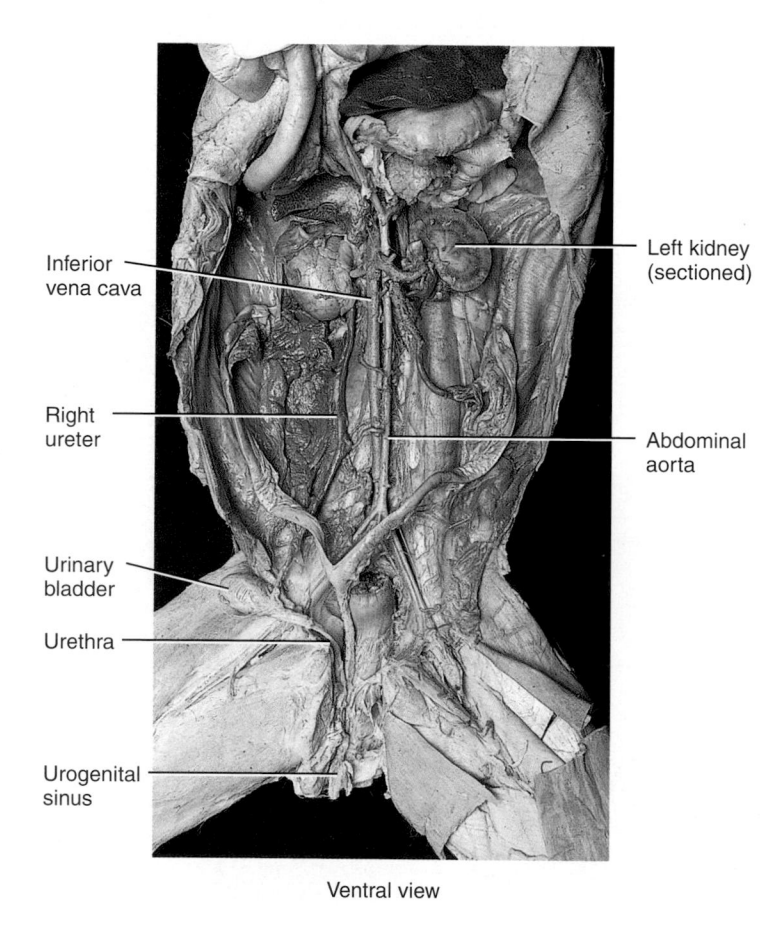

Inferior
vena cava

Right
ureter

Urinary
bladder

Urethra

Urogenital
sinus

Left kidney
(sectioned)

Abdominal
aorta

Ventral view

FIGURE A.17B
Female urinary system

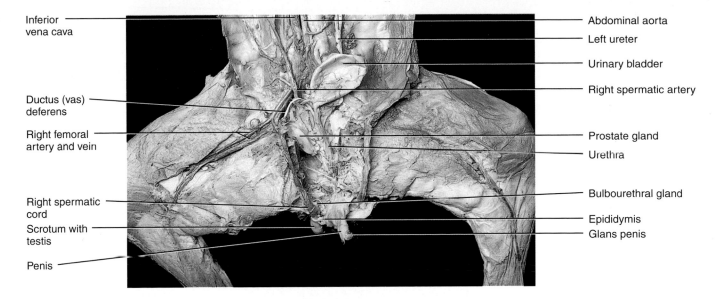

FIGURE A.18A
Male reproductive system

Inferior vena cava

Ductus (vas) deferens

Right femoral artery and vein

Right spermatic cord

Scrotum with testis

Penis

Abdominal aorta

Left ureter

Urinary bladder

Right spermatic artery

Prostate gland

Urethra

Bulbourethral gland

Epididymis

Glans penis

Ventral view

FIGURE A.18B
Female reproductive system

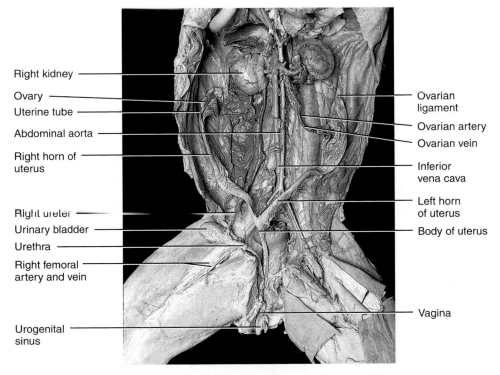

Right kidney

Ovary

Uterine tube

Abdominal aorta

Right horn of uterus

Right ureter

Urinary bladder

Urethra

Right femoral artery and vein

Urogenital sinus

Ovarian ligament

Ovarian artery

Ovarian vein

Inferior vena cava

Left horn of uterus

Body of uterus

Vagina

Ventral view

Illustration Credits

The following illustrations are adapted from *Principles of Human Anatomy* 8th edition by Gerard Tortora (New York, 1999) John Wiley & Sons ©1999 Biological Sciences Textbooks, Inc.:

1.1, 1.2, 1.3, 1.4, 1.5, 2.1, 2.2, 2.3, 2.4, Table 3.1, 3.2, 3.3, 3.4, 3.6, 3.8, 3.9a-b, 3.10a, 3.11, 3.16, 3.17a-c, 3.18, 3.20, 3.21b, 3.22a, 3.23, 3.24, 3.25, 3.26, 3.27, 3.28, 3.29, 3.30, Table 4.1, 4.4, 4.5, 4.8, 5.1, 5.2, 5.4, 5.5, 5.7, 5.8, 5.11, 5.12, 5.13, 5.16, 5.17, 5.18, 5.19, 6.1, 6.2a-b, 6.3, 6.4, 6.5, 6.7, 6.13, 6.15, 6.17, 6.19, 6.23, 7.1, 7.2, 7.3, 7.4, 8.1, 8.2, 8.3, 8.4, 8.5, 8.8, 8.13, 8.14, 8.16, 8.18, 8.19, 8.21, 8.22, 8.25, 8.26, 9.1b-c, 9.2a-d, 9.3a, 9.3c, 9.4, 10.1, 10.2, 10.3, 10.4, 10.5, 10.6, 11.1, 11.3, 11.4, 11.7, 11.9, 11.10, 11.11, 11.13, 11.14, 11.15, 12.1, 12.5, 12.6, 12.7, 12.8b, 12.9b-c, 12.11, 12.12, 12.13b, 13.1, 13.3, 13.4, 13.5, 13.7, 13.8, 13.9, 14.1, 14.3, 14.4, 14.5, 14.8, 14.9, 14.10, 14.11b, 14.12, 14.13, 14.14a, 15.1, 15.2

The following illustrations are adapted from *Principles of Anatomy and Physiology* 8th edition by Gerard Tortora and Sandra Reynolds Grabowski (New York: John Wiley & Sons, 1996) ©1996 Biological Sciences Textbooks, Inc. and Sandra Reynolds Grabowski: 4.2, 9.7, 12.4

3.1a, 3.31 From *Atlas of the Human Skeleton* by Gerard J. Tortora (New York, 1996) John Wiley & Sons.

14.14b From *Introduction to the Human Body*, 4th edition by Gerard J. Tortora ((New York: John Wiley & Sons, 1997) ©1997 Biological Sciences Textbooks, Inc.

Molly Borman: 7.1, 9.2, 11.1, 12.6.
Leonard Dank: 5.1.
Steve Oh: 10.1, 12.2.
Lynne O'Kelley: 1 .4
Nadine Sokol: Table 3.1.
Kevin Somerville: 1.1-1.3, 1.5, 6.1, 8.1, 12.8, 13.1, 14.1, 14.4, 14.13.
Kevin Somerville/Nadine Sokol: 2.1-2.3.

Photo Credits

Chapter 1

Figure 1-5a: From Stephen A. Kieffer and E. Robert Heitzman, *An Atlas of Cross-Sectional Anatomy,* Harper & Row, Publishers, Inc., New York. Figure 1-5b: Lester V. Bergman/Project Masters, Inc. Figure 1-5c: Martin Rotker.

Chapter 2

Figure 2-1a: Biophoto Associates/Photo Researchers. Figures 2-1b-d,f-i and k: Courtesy Michael Ross, University of Florida. Figure 2-1e: Biophoto Associates/Photo Researchers. Figure 2-1j: Lester V. Bergman/Project Masters, Inc. Figure 2-2a-e,h-k: Courtesy Michael Ross, University of Florida. Figure 2-2f: Courtesy Andrew J. Kuntzman. Figure 2-2g: Ed Reschke. Figure 2-2l: John Burbidge/Photo Researchers. Figure 2-2m (top): Courtesy Michael Ross, University of Florida. Figure 2-2m (bottom): John D. Cunningham/Visuals Unlimited. Figure 2-3a,b: Courtesy Michael Ross, University of Florida. Figure 2-3c: Biophoto Associates/Photo Researchers. Figure 2-4: Ed Reschke.

Chapter 3

Chapter photographs: Mark Nielsen.

Chapter 4

Chapter photographs: Mark Nielsen.

Chapter 5

Chapter photographs: Mark Nielsen

Chapter 6

Figure 6-2a: Courtesy Andrew J. Kuntzman. Figure 6-2b: Courtesy Dr. Dennis Strete. Figure 6-2c: Courtesy Michael Ross, University of Florida. Figures 6-3- 6-23: Mark Nielsen.

Chapter 7

Figure 7-2a: From *The Color Atlas of Anatomy*, 3e., by J.W. Rohen and C. Yokochi. Reprinted with permission of Wiliams and Wilkins. Figures 7-2b,c: Courtesy Michael Ross, University of Florida. Figures 7-3, 7-4: Mark Nielsen.

Chapter 8

Figures 8-2- 8-4, 8-6- 8-16, 8-18- 8-20 and 8-23-8-25: Mark Nielsen. Figure 8-5: Jean Claude Revy/Phototake. Figure 8-17: From N. Gluhbegovic and T.H. Williams, *The Human Brain: A Photographic Guide*, Harper and Row Publishers, 1980. Figures 8-21 and 8-22: Stephen A. Kieffer and E. Robert Heitzman, *An Atlas of Cross-Sectional Anatomy*, Harper & Row Publishers, New York. Figure 8-26a: Jan Leesma/Custom Medical Stock Photo. Figure 8-26b: From Richard G. Kessel and Randy H. Kardon, *Tissues and Organs: A Text Atlas of Scanning Electron Microscopy*. Copyright ©1979 by W. H. Freeman and Company. Reprinted by permission. Figure 8-26c: Dennis Kunkel/Phototake. Figure 8-26d: ©Fujita.

Chapter 9

Figure 9-1a: Courtesy Lynne Marie Barghesi. Figure 9-1b: Mark Nielsen. Figure 9-1c: Courtesy Michael Ross, University of Florida. Figure 9-2a: From Yokochi, Rohen, Weinreb, *Photographic Anatomy of the Human Body*, 3e. Tokyo: Igaku-Shoin, 1989. ©1989 by Igaku-Shoin, Ltd.

Figures 9-2b and 9-2c: From Richard G. Kessel and Randy H. Kardon,*Tissues and Organs: A Text Atlas of Scanning Electron Microscopy* Copyright ©1979 by W. H. Freeman and Company. Reprinted by permission. Figure 9-2d: Courtesy Michael Ross, University of Florida. Figure 9-3a: ©John Wiley & Sons. Figure 9-3b: Courtesy Michael Ross, University of Florida. Figure 9-3c: Dr. Dennis Strete. Figure 9-3d: Robert Chase. Figure 9-4a: ©John Wiley & Sons. Figure 9-4b: John D. Cunningham/Visuals Unlimited.

Chapter 10

Figures 10-2a, 10-3a, 10-4a, 10-5a, 10-6a: Mark Nielsen. Figure 10-2b: Courtesy Dr. Dennis Strete. Figures 10-3b, 10-5b: Courtesy Michael Ross, University of Florida. Figure 10-6b: CNRI/Phototake.

Chapter 11

Figures 11-2- 11-8 and 11-12- 11-14: Mark Nielsen. Figure 11-9a: John D. Cunningham/Visuals Unlimited. Figure 11-9b: Courtesy Andrew J. Kuntzman. Figures 11-9c and 11-10: Courtesy Michael Ross, University of Florida. Figure 11-11: From J. W. Rohen, Ch. Yokochi, E. Luetjen, Drecoll, *Color Atlas of Anatomy*, 5e., Lippincott Williams & Wilkins Publishers. Figure 11-15a: Biological Photo Service/Photo Researchers. Figure 11-15b: Biophoto Associates/Photo Researchers.

Chapter 12

Figures 12-2, 12-3, 12-5, 12-6a, 12-8a, 12-9a-b, 12-10, 12-11a, 12-12b, 12-13a: Mark Nielsen. Figure 12-4: From Stephen A. Kieffer and E. Robert Heitzman, *An Atlas of Cross-Sectional Anatomy*, Harper & Row, Publishers, Inc., New York. Figure 12-6b: Courtesy Michael Ross, University of Florida. Figure 12-7a: Dr. D. Gentry Steele. Figure 12-7b: H. Hubatka/Maritus Gnbj/Phototake. Figure 12 8b: Courtesy Michael Ross, University of Florida. Figure 12-9c: Ed Reschke. Figure 12-11b: Stephen A. Kieffer and E. Robert Heitzman, *An Atlas of Cross-sectional Anatomy*, Harper & Row, New York. Figure 12-11c: Courtesy Michael Ross, University of Florida. Figure 12-12a: From David L. Bassett, *A Stereoscopic Atlas of Human Anatomy*. Reproduced with permission. Figure 12-12c: Fred E. Hossler/Visuals Unlimited. Figure 12-12d: Willis/Biological Photo Service. Figure 12-13b: Roland Birke/Peter Arnold, Inc. Figure 12-13c: Courtesy Michael Ross, University of Florida.

Chapter 13

Figures 13-2- 13-6 and 13-9a: Mark Nielsen. Figure 13-7: Dr. Dennis Strete. Figure 13-8 and 13-9b: Biophoto Associates/Photo Researchers.

Chapter 14

Figures 14-2, 14-3a-b: Mark Nielsen. Figure 14-3c: Ed Reschke. Figure 14-4: Courtesy Michael Ross, University of Florida. Figure 14-5a,b: Michler/Science Photo Library/Photo Researchers. Figure 14-6: Mark Nielsen. Figure 14-7: From Yokochi, Rohen, Weinreb, *Photographic Anatomy of the Human Body*, 3e. Tokyo: Igaku-Shoin, 1989. ©1989 by Igaku-Shoin, Ltd. Figures 14-8 and 14-9: Mark Nielsen. Figure 14-10a,b: Biophoto Associates/Photo Researchers. Figure 14-11a,b: Courtesy Michael Ross, University of Florida. Figure 14-12: John D. Cunningham/Visuals Unlimited. Figure 14-13: Biophoto Associates/Photo Researchers. Figure 14-14a: Mark Nielsen. Figure 14-14b: Courtesy Michael Ross, University of Florida.

Chapter 15

Chapter photographs: ©John Wiley & Sons.

Chapter 16

Figure 16-1a: Photo provided courtesy of Kohei Shiota, Congenital Anomaly Research Center, Kyoto University, School of Medicine, Japan. Figure 16-1b-e: Courtesy National Museum of Health and Medicine, Armed Forces Institute of Pathology. Figure 16-1f: ©Lennart Nilsson, from *A Child Is Born*. Figure 16-1g,h: Photo provided courtesy of Kohei Shiota, Congenital Anomaly Research Center, Kyoto University, Japan.

Appendix

Chapter photographs: ©John Wiley & Sons.

Index